NASA
MERCURY

1956 to 1963 (all models)

To the Mercury Seven:

Alan B Shepard (1923-1998)
Virgil I Grissom (1926-1967)
John H Glenn (1921-2016)
Malcolm Scott Carpenter (1925-2013)
Walter M Schirra (1923-2007)
L Gordon Cooper (1927-2004)
Donald K Slayton (1924-1993)

First published in May 2017
Reprinted in November 2017, January 2020, 2021 (twice) and 2024

A catalogue record for this book is available from the British Library.

ISBN 978 1 78521 064 8

Library of Congress control no. 2016959362

Published by Haynes Group Limited,
Sparkford, Yeovil,
Somerset BA22 7JJ, UK.
Tel: 01963 440635
Int. tel: +44 1963 440635
Website: www.haynes.com

Haynes North America Inc.,
2801 Townsgate Road, Suite 340,
Thousand Oaks, CA 91361

Printed in India.

NASA
MERCURY

1956 to 1963 (all models)

Owners' Workshop Manual

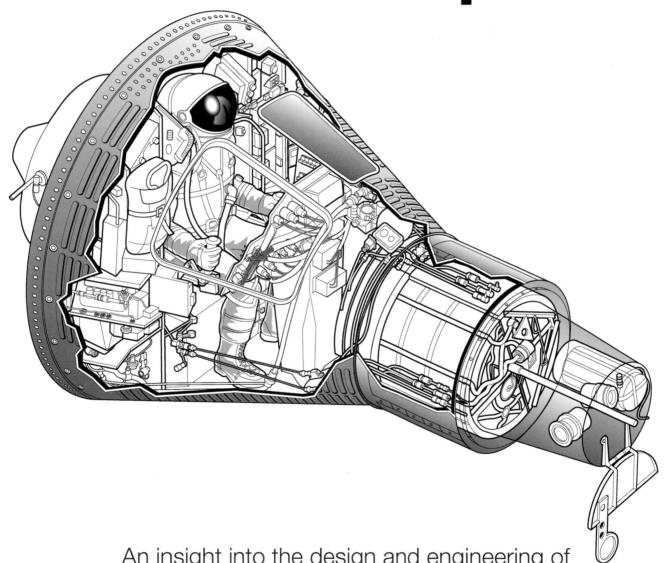

An insight into the design and engineering of
Project Mercury – America's first manned space programme

David Baker

The memorial at Cape Canaveral to the seven Mercury astronauts.
(via David Baker)

Contents

DK Slayton (signature)

Alan Shepard (signature)

**FREEDOM
7**

Astronaut
Alan B. Shepard
Suborbital Flight
5 May 1961

Scott Carpenter (signature)

AURORA

Astronaut
M. Scott Carpenter
Three Orbits
24 May 1962

Gus Grissom (signature)

**LIBERTY
BELL
7**

Astronaut
Virgil I. Grissom
Suborbital Flight
21 July 1961

Astronaut
Walter M. Schirra
Six Orbits
3 October 1962

Wally Schirra (signature)

Gordon Cooper (signature)

JH Glenn, Jr. (signature)

Friendship 7

Astronaut
John H. Glenn
Three Orbits
20 February 1962

**FAITH
7**

Astronaut
L. Gordon Cooper
15-16 May 1963
22 Orbits

MERCURY SPACECRAFT

Designed and built by MCDONNELL, St. Louis

for the National Aeronautics and Space Administration

Introduction

The first American spacecraft designed to carry humans into Earth orbit was conceived by the US Air Force, developed by NASA and built by industry. Originating as a quick means of obtaining early physiological data on the performance of the human body in space, what would become the Mercury spacecraft was closely linked to the first attempt by the US Air Force to establish a programme involving spaceplanes, powerful rockets, orbiting manned vehicles and bases on the Moon.

Across the history of the space programme, the US Air Force has three times tried to establish a human presence in space and on each occasion it has failed to secure necessary funding and political approval. The first attempt was when its plan for getting the first man in orbit was transferred to the newly formed NASA in 1958, the second when its plan for an orbiting spaceplane (Dyna-Soar) was cancelled by the Kennedy administration in 1963, and the third when the proposed Manned Orbiting Laboratory was cancelled by the Johnson administration six years later.

Because the US Air Force had such an important role in bringing to NASA – when it was formed in October 1958 – the conceptual plan for what would become the Mercury spacecraft, that part of the story is important, if less well known. For this reason the first chapter details the origin and development of plans for a military manned presence in space.

Brought in at first as consultants on the Air Force plan for a Man-In-Space-Soonest (MISS) capsule, the National Advisory Committee for Aeronautics (NACA) was given responsibility for the programme, renamed the National Aeronautics and Space Administration, effective 1 October 1958. The tangled story of how all that happened is the subject of the first chapter. It is a vital prelude to the more familiar story of the Project Mercury programme and the technical description that ensues.

OPPOSITE Built by McDonnell Aircraft Company and operated by NASA, the Mercury spacecraft was flown by six astronauts, four of whom became the first Americans to orbit the Earth outside the atmosphere. *(McDonnell)*

BELOW Aboard Freedom 7 for his ballistic flight to space and back, Alan Shepard anticipates his impending flight as America's first astronaut. *(NASA)*

Chapter One

Origins (1945–58)

The story of the Mercury spacecraft begins in an Air Force spaceplane project of the mid-1950s called Dyna-Soar, an evolved concept with an offensive capability, including the delivery of thermonuclear (radiation pressure) weapons to any point on Earth. It was not planned for Dyna-Soar to remain in Earth orbit for extensive periods but to traverse space in order to reach hemispheric targets quickly and at short notice.

OPPOSITE Designated 'shape B', this early 1958 design satisfied the essential requirements set by the Air Force for its Man-In-Space-Soonest (MISS) concept. *(USAF)*

ABOVE By the early 1950s the National Advisory Committee for Aeronautics (NACA) had decided on the next level of high-speed flight research by investigating hypersonic flight beyond Mach 5. This in turn led to the development of the North American X-15, a precursor to plans for ballistic flights into space to provide information on the physiological reactions to humans in weightlessness. In this view an X-15 is carried aloft under the starboard wing of a modified Boeing B-52, with a North American F-100 flying 'chase'. *(NASA)*

RIGHT Test pilot Milton O. Thompson holds a model of the basic X-15 (black) and the modified X-15A-2 that explored the outer reaches of the aircraft's performance envelope, reaching Mach 6.7 in October 1967. *(NASA)*

Dyna-Soar was a contraction of 'dynamic soaring', a euphemism for what would later be called trans-atmospheric vehicles (TAVs) – manned or unmanned hypersonic strike systems, exiting the atmosphere to enter orbit at 17,500mph (28,100kph). After traversing an appropriate arc of the first orbit the dart-shaped Dyna-Soar would slow down using retro-rockets, deploy its weapons from the stratosphere to a pre-selected target and conduct 'dynamic soaring' using high lift-over-drag atmospheric flight to return to base at hypersonic speed, having completed one full circuit of the Earth.

Dyna-Soar was never built but in the 1950s, when it was a highly logical progression for faster and higher flight, the Air Force needed diagnostic information about the physiological responses of a pilot to such extremes. To get this information

they devised a step programme to ease the way in to the hypersonic world of dynamic soaring air vehicles flying through space.

Phase One would require a preliminary analysis of pilots exposed to the accelerations of a rocket, weightlessness in space and the decelerations of coming back down through the atmosphere. This would be achieved, said the Air Force, through a ballistic rocket flight gathering medical data for refinement of the design and operations planning for the Dyna-Soar spaceplane. That would settle questions about how well humans could work at piloting tasks under such conditions.

Phase Two would be Dyna-Soar itself, proceeding to operational capability through a series of separate stages. Launched by a ballistic missile, Dyna-Soar Step I was to be a technology demonstrator for evaluating the concept and the engineering, much as Phase One had qualified humans for the job of flying such a craft. It would have a range of 5,000 miles (8,045km) and provide much needed data on the aero-thermal characteristics of the outer atmosphere.

Dyna-Soar Step II would be a developed version to conduct reconnaissance while expanding the flight regime to give the spaceplane a range of more than 10,000 miles (16,090km). The fully operational role of strategic bombardment would be the responsibility of Dyna-Soar Step III, entering orbit and providing a strategic offensive role with a truly global range.

The Air Force was exceedingly conservative about the schedule, not expecting Dyna-Soar Step I to become operational before 1963, Dyna-Soar II in 1968 and Dyna-Soar III in 1974. There was good reason for the cautious approach. In the mid-1950s the Air Force was involved with the development of the X-15, a hypersonic research aircraft that aimed to investigate the flight regime beyond the capabilities of existing vehicles. At this time no piloted aircraft had exceeded Mach 3 and the X-15 was expected to investigate the flight regime at Mach 5 and above.

At Mach 5 the X-15 would have only 4% of the kinetic energy required to achieve orbital flight, defined by the square of the vehicle's Mach number ratio as an approximation of the velocity ratio. Clearly, there was probably a lot

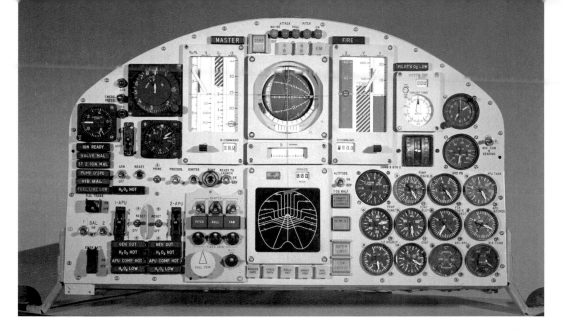

of unknown physics lurking in that remaining 96% and the phased approach to achieving the full requirement for orbital flight with Dyna-Soar Step III was the most prudent way to proceed.

The key to piloted hypersonic flight was research into the conditions that aerospace vehicles would encounter at very high altitude and speeds in excess of Mach 5, considered the threshold of hypersonic flight. Even before the X-15 was drafted high-altitude research programmes were conducted as a means of investigating the relatively unknown environment in this new flight regime.

It was this research that eventually got the attention of the National Advisory Committee for Aeronautics (NACA), the precursor of NASA, and its Pilotless Aircraft Research Division (PARD) located at Wallops Island, Virginia. In 1953 they began conducting a series of experimental flights using small multistage research rockets carrying packages of instruments to the upper atmosphere and to hypersonic velocities. On 20 November 1953 PARD flew a parabolic-shaped nose section to Mach 5.0 to measure the heat transfer characteristics of the Inconel material from which it was fabricated.

A lot changed between 1953 and 1956: the X-15 was defined and design began; a decision was taken to proceed with spy satellites; the United States began to field miniaturised thermonuclear weapons; and, taking advantage of the reduced size and weight of warheads, Intercontinental Ballistic Missiles (ICBMs) were authorised and contractors began to design what would become Atlas and Titan rockets. Moreover,

North American Aviation was now building the X-15, and Dyna-Soar would follow as hypersonic flight results fed into the boost-glide programme.

A task group of the source selection board had 111 contractors lined up to bid on the various elements of research required to make a decision about the type of vehicle required for each evolutionary step, with emphasis on the initial two-year research phase. By March 1958 nine contractor teams had submitted proposals for the first step in the Dyna-Soar programme but each had a distinctly unique idea as to how the objective should be met. Evaluation was completed during April, and on 16 June 1958 Martin and Boeing were selected to develop separate bids for Dyna-Soar I. Following a period of up to 18 months a single contractor would be chosen.

Power game

Several factors were now in play that would significantly shift the balance of power in getting a man into space. On 4 October 1957 the Russians had orbited the world's first artificial satellite, Sputnik 1, sending shock waves through defence and intelligence establishments throughout the Western world. In America the response was immediate. After the attempted launch of a Vanguard satellite from Cape Canaveral failed on 6 December, President Eisenhower authorised the Army Ballistic Missile Agency (ABMA) to hurriedly

prepare a military research rocket, Jupiter-C, to send up an instrumented package in response. That was successfully accomplished on 31 January 1958.

The repercussions of Sputnik in the military were most immediately evident when Secretary of Defense Neil McElroy, acting on the orders of President Eisenhower, set up the Advanced Research Projects Agency (ARPA) on 7 February 1958 as the fifth US military service (after the Air Force, the Army, the Navy and the Marine Corps). It had authority to collect and direct research programmes for all the other services and was based in the Pentagon. It was the access gate through which all future technology programmes would have to pass. Within a few weeks all US space programmes were placed with ARPA.

A peripheral consequence of Sputnik was that 'space' – gathering transient, transitory and esoteric concepts which never previously had full legitimacy – was suddenly respectable and a have-to-have adjunct to military thinking. But it was the politicians who needed persuading that this was the case; the scientists, the engineers

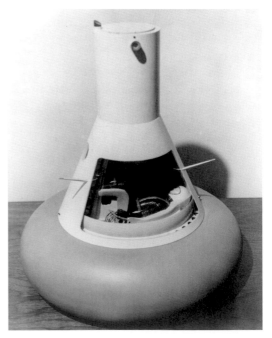

BELOW An alternative configuration envisaged this toroidal landing impact skirt, inflated after re-entry to absorb the shock of impact on water, which was considered necessary for an over-ocean launch from Cape Canaveral. *(USAF)*

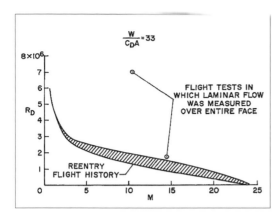

NOSE SHAPE	⊕	R=⅓d 15°	R=2d 53°	R=1.5d
W/c_DA	57	104	37	33
TOTAL HEAT INPUT, BTU	338,000	414,000	293,000	256,000
MAXIMUM HEATING RATE AT STAG POINT BTU/(SQ FT) (SEC)	125	215	166	64.5
c_{N_α}	.92	.97	.69	.24
c_{L_α}	0	.47	-.71	-1.32
c_{m_α}	-.16	-.09	-.22	-.13

and the military officials responsible for the boost-glide concept and a multitude of space-related concepts had few doubts.

When Dyna-Soar emerged as the chosen boost-glide programme the National Advisory Committee for Aeronautics was approached by the Air Force for consultative technology advice. On 31 January 1958, the very day the US Army launched America's first satellite, Explorer 1, General Donald R. Putt, USAF director-general for development, spoke to NACA about Dyna-Soar I running much like the X-15 programme and implied that he would be happy to have it managed by this civilian, government-run research agency. But the Air Research and Development Command (ARDC) opposed that, since the boost-glide vehicle was intended to mature into a full weapon system and should, they said, be retained fully within the control of the Air Force. Nevertheless, NACA was encouraged to assist in a support role, using its laboratories to carry out essential basic research and testing and to keep the Air Force advised on flight results from the X-15, when it started flying.

Late of General Electric, Roy W. Johnson was

appointed to head ARPA and he let it be known publicly that it had 'long term development responsibility for manned space flight as soon as technology permits'. The Air Force had been pressing for some time to have an early exposure of a pilot to the physiological stresses of space flight in order that results could inform the human and ergonomic elements of space-vehicle design as they applied to Dyna-Soar. That work had also involved NACA.

MISS

As the essential prerequisite to boost-glide flight, on 15 February 1956 the ARDC commander General Thomas Power had declared at a staff meeting the need to get a man in orbit as quickly as possible. The ARDC pushed hard for a dual approach: long-term development of the boost-glide vehicle preceded by a convergence of the requirement for bio-physiological data with a simplified ballistic 'capsule' giving the Air Force experience with the space environment.

To some degree this 'quick and dirty'

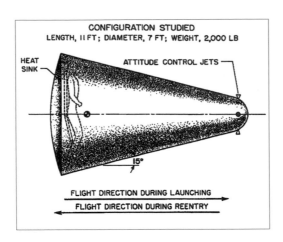

CONFIGURATION STUDIED
LENGTH, 11 FT; DIAMETER, 7 FT; WEIGHT, 2,000 LB

HEAT SINK

ATTITUDE CONTROL JETS

15°

FLIGHT DIRECTION DURING LAUNCHING

FLIGHT DIRECTION DURING REENTRY

RIGHT Using the preceding evaluations and analyses, a specific configuration was chosen for the Air Force MISS concept incorporating a low-drag nose for forward flight up through the atmosphere and a high-drag heat-resistant shield for re-entering. This logically dictated the supine position of the occupant together with attitude control thrusters already proven on other designs including the X-15, still in development at this date. *(NACA)*

approach of a man-carrying capsule ran counter to NACA's measured, scientific approach which preferred the concept of lifting-vehicles possessing an aerodynamic shape to minimise drag and execute controlled landings. But at the behest of Powers, a small group of staffers prepared a strategic plan involving a 'Manned Glide Rocket Research System' preceded by a 'Manned Ballistic Rocket Research System'. The latter would comprise a detachable nose cone launched by an ICBM that could put a manned capsule in orbit. From this, said the staffers, everything else would flow.

While most of the scientists and engineers working with the Pilotless Aircraft Research Division at NACA-Langley wanted the value of a lifting glide vehicle, they were persuaded by the Air Force to study non-lifting ballistic capsules as a quick means of getting people in orbit for physiological research. And for that, they proposed a spherical non-lifting capsule.

During the rest of the year NACA worked with several contractors on low-level studies to determine how that could be achieved. The McDonnell Aircraft Company started work on preliminary design concepts in March 1956 and Avco applied its detailed work on missile re-entry warheads to the challenge, delivering its first conclusions later that year.

Atlas was the only missile under development capable of lifting a manned capsule into orbit, albeit one weighing only one tonne, and it was already struggling with a range of technical problems. Outside the 'need to know' group, NACA engineers were not given access to the detailed specifications of Atlas and therefore had little to go on when designing a capsule. In effect they were just consultants to the Air Force,

which had little incentive to share technical details when they were themselves striving to be dominant in the field.

But several technical developments had taken place that would favour the ballistic capsule over the boost-glide concept, and while NACA was naturally inclined to produce aerodynamic concepts some – particularly at Langley – became increasingly convinced that a non-lifting capsule would be the optimum way to get men into space. Despite this, in a feasibility study issued in September 1957 NACA officially supported the Manned Glide Rocket Research System with a shaped lifting-body re-entry vehicle. It was this configuration that morphed into the Dyna-Soar concept agreed at a technical conference on 15 October, selected by ARPA five days earlier.

On 8 March 1958 the Air Force Ballistic Missile Division (BMD) laid out a highly ambitious series of objectives up to and including manned lunar landing and return. In the months after the Sputnik launches outer space had become a destination rather than an application and talk about boost-glide vehicles and ballistic capsules merged into a more generalised desire to put the Air Force in space and to keep it there. Two days later the BMD hosted a major conference at its Los Angeles headquarters where specialists and representatives from government and industry heard how the Moon goal was the end objective but that a manned orbital flight around Earth was a priority.

Leading the push to achieve this as quickly as possible was General Bernard A. Schriever, the driving force behind the expanded ballistic missile programme that had dominated strategic planning since 1954. He was now running the Atlas and Titan ICBM programmes. Schriever was convinced that leadership in rockets, missiles and space was the natural prerogative of the Air Force and that future activity in those fields would be managed by the Ballistic Missile Division.

Conceptual development of the MISS programme began to accelerate. At the ARDC conference at the Ballistic Missile Division on 10–12 March 1958, where 80 representatives of industry embracing both engineering and aeromedical sciences met, the basic specification was discussed. They agreed that the capsule would have a diameter of 5.0ft (1.8m) and a length of 7.9ft (2.4m), weigh no more than 3,970lb (1,800kg) and return to a splashdown in water. It would be capable of supporting flights in orbit of up to two days, a duration set by the technical capability of the life support system.

On 14 March the ARDC delivered its plan for a manned ballistic capsule and on 19 March the Air Force under-secretary asked ARPA to fund the development of this $133 million project. For much of April the ARDC Man-In-Space Task Force worked at the Ballistic Missile Division, contributing to the Air Force Manned Military Space Development Plan. Enshrined was its ultimate goal of working to 'land a man on the Moon and return him safely to Earth', language almost identical to a phrase used by President Kennedy at his special address to the nation on 25 May 1961 when ordering NASA to send astronauts to the Moon.

The first phase of this Air Force plan was Man-In-Space-Soonest (MISS), utilising a ballistic spacecraft that would first carry primates and then a human pilot. The second phase was Man-In-Space-Sophisticated which would support a two-man crew on Earth-orbiting flights of up to 14 days. This would be followed by Lunar Reconnaissance, a period of scientific probing of the Moon with instrumented spacecraft, leading to Manned Lunar Landing and Return, first with primates, then with men orbited around the Moon before a landing.

But there was pressure from General LeMay, who wanted a much faster pace than the proposed MISS programme would make possible, and he brandished the results of a study by Convair and Avco in which a very simple capsule would be launched on an Atlas rocket. BMD came up with a plan using a basic Atlas rocket to send up a manned capsule weighing no more than 3,000lb (1,360kg) in April 1960 for $99.3 million. On 16 June 1958 ARPA approved the MISS plan and authorised BMD to proceed on that basis. The Wright Air Development Center put North American Aviation and General Electric to work designing the capsule and its environmental control system together with the production of associated mock-ups and working models of the interior.

But ARPA was reluctant to release the

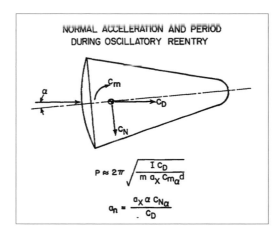

full funding required and held a meeting with several companies and NACA in late June to discuss technical issues, with little agreement. But the Air Force was not only struggling to persuade ARPA to get behind MISS and release full funding. It was also fighting off pressure from the Army in a feverish frenzy of activity that saw the Pentagon besieged with ideas about putting a man in space.

The Army had Wernher von Braun, the charismatic V-2 rocket engineer who had been working for them since 1946. Now at Redstone Arsenal, his team had been responsible for the Redstone missile, largely based around the V-2 but with several improvements and a much more efficient rocket motor. Late in 1957 Von Braun came up with the idea – initially called Man Very High, and then Project Adam – of putting a detachable capsule on top of a Redstone rocket and launching it and its pilot on a ballistic hop to an altitude of 170 miles (273km) before recovering them from the

METHODS FOR PRODUCING LIFT	
CONTROL METHOD	PERFORMANCE
REACTION JET	$L = 12\ T_j$
DRAG FLAP $\frac{S_f}{A} = 0.045$	$M = 6.9$
C.G. SHIFT Δy	$\frac{L}{D} = 10.1\ \frac{\Delta y}{d}$

LEFT This figure shows the level of oscillation and that the length of the period (p) is directly proportional to the square root of the centre of gravity divided by the pitching-moment coefficient slope per radian, based on the maximum diameter. Similarly, the severity of oscillation can be decreased by reducing the resulting normal acceleration for a given angle of attack. From this it can be shown that a flat nose is favourable. (NACA)

LEFT To simplify recovery, the Air Force and NACA examined various shapes that could be used to produce lift for terminal guidance and unpowered directional control. Because of the low value of the pitching-moment coefficient slope per radian, a high moment arm and a high negative centre of lift, thrusters would be good for this. A drag flap was shown to be effective during tests at the Langley Laboratory's 11in hypersonic tunnel. A small centre-of-gravity shift would also produce useful results. (NACA)

RIGHT Taking all these factors into consideration, several shapes were tested using the basic cone-shaped configuration. Here Shape A presents a highly pointed exit configuration with retro-rockets beneath the forebody (the base of the cone), with the possibility of jettisoning the nose to present a bluff afterbody. *(NACA)*

BELOW Calculations on the level of thermal load during high-speed re-entry revealed a prediction for latent thermal soak, and these derivations were validated with the flight of Big Joe on 9 September 1959. *(NASA)*

Atlantic Ocean. This would, said the Army, be much quicker and less costly than MISS and could achieve its objective during 1959.

Ironically, in the UK, during 1946 R.A. Smith of the British Interplanetary Society had proposed an identical application for the German V-2 ballistic missile. Under Project Megaroc, Smith suggested that a capsule carried on a ballistic trajectory to space on top of a V-2 would provide early research results on the physiological reaction of humans to space flight. The British had been the first to fire captured V-2 rockets immediately after the war as an evaluation of the technology. Despite

it being a sound engineering possibility, the economic state of Britain immediately after the war prevented Megaroc from being approved. If it had, a British astronaut could have been the first person in space, probably in the late 1940s. But that was not to be.

A derivative of the Redstone, called Jupiter-C, had already been used to send the first American satellite into orbit on 31 January and was being developed into a general satellite launcher as Juno I. There was precedent for believing it could start the process of understanding how humans reacted to space flight and the unknowns of weightlessness. But NACA's boss Hugh Dryden was contemptuous of this approach, claiming that it had 'about the same technical value as the circus stunt of shooting a young lady from a cannon'. On 11 July 1958 Roy Johnson rejected the idea. Ironically, it was precisely this application that would carry, twice, the first Americans into space as part of Project Mercury three years later.

End game

The last major action in support of the Air Force MISS capsule, designated Project 7969, was the preliminary selection finalised on 25 June 1958 of the initial tranche of pilots who would be carried into space: Neil Armstrong, William Bridgeman, Scott Crossfield, Iven Kincheloe, John McKay, Robert Rushworth, Robert Walker and Alvin White. These men were the world's first astronaut selection.

When NASA took over the MISS project these appointments were nullified and the space agency would make its own selection starting in early 1959. A new Air Force selection, for Dyna-Soar pilots, would be made in secret during 1960 and some of these names would appear there too. It is interesting to note that Armstrong was the only pilot selected for all three: MISS, Dyna-Soar and as a NASA astronaut (in September 1962).

The events which wrested control of plans to send Americans into space from the Air Force were the launch of the world's first artificial satellite, Sputnik 1 on 4 October 1957, and the first animal sent to orbit the Earth, the dog Laika in Sputnik 2, less than a month later. In the United States there was a universal belief that the Soviet Union was developing an aggressive

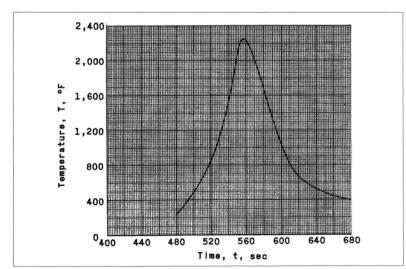

and fast-paced programme of ballistic missile deployment threatening North America and its allies with nuclear attack, and to serve as a robust and convincing demonstration to politically uncommitted countries that Russia was the world power of the future.

The legislature in Washington DC held a rapidly convened series of hearings during which the future path for US activities in space was deliberated. Busloads of experts, specialists, eminent scientists, technologists, industrial magnates and lobbyists were processed through these hearings in the nation's capital with one purpose: to find a way to respond to the unexpected achievements of the Soviet Union. For its part the White House, and the executive arm of government, cautioned against overreaction, with the President enforcing his belief that the scientific exploration of space should be conducted by a civilian government agency.

That decision was made by the President on 5 March 1958, after Eisenhower had set up the office of Special Assistant to the President for Science and Technology, appointing James R. Killian to head it, and reconstituted the President's Scientific Advisory Committee. Further US satellite launches were planned using both the Army's Jupiter-C rocket and the Vanguard rocket which on 15 March orbited its first satellite, but representation from the scientific community – epitomised by the National Academy of Sciences – endorsed the White House decision.

Eisenhower had consulted his Advisory Committee on Government Organization, which eased through the recommendation that a transformed NACA would be the most appropriate body to gather up all the non-military space activities and rename it the National Aeronautics and Space Administration (NASA). The memorandum outlining this shift referred to the general lack of competence on space matters in NACA, noting that almost all space-related activity had been conducted by the Department of Defense. It advised the new organisation to 'tap into this competence without impairing the military space program'.

The Administrator of NASA was to be selected by the President and the composition of the former NACA committee was to be changed. The memo also accepted that the

Act enabling these steps to be taken should be expedited quickly and that the path should be laid for NASA to induct organisations previously a part of the military for whose work there was no immediate military requirement. This alone would open the door for the majority of the rocket team at Redstone Arsenal to move, eventually, to NASA and be renamed as the Marshall Space Flight Center. In terminology too, the 'laboratories' would be renamed 'research centers'.

The legislation was sent to Congress on 2 April but both houses had already acted to provide the essential bodies necessary to deal with space programmes and all the decision-making that this implied. The Senate set up the Special Committee on Space and Astronautics with Lyndon Johnson as its chair, while the House created the Select Committee on Astronautics and Space Exploration with John McCormack, majority leader, as chair. Hearings on the National Aeronautics and Space Act began on 15 April and it was passed on 16 July, the President signing it 13 days later.

Before that Act had been formally signed, Dr Hugh Dryden, director of NACA, sent a letter to James Killian on 18 July asserting his claim to Man-In-Space-Soonest so as to clear the air between the civilian agency and the Air Force. But there was an air of uncertainty. There was no official directive as to which organ of government should be responsible for manned space flight. Neither the outgoing NACA nor the incoming NASA had moral authority to the project and it had been conceptualised, created

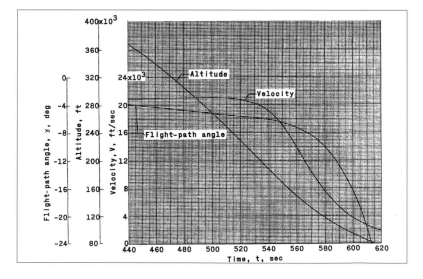

BELOW Re-entry parameters for the Big Joe flight corroborated calculations and theoretical predictions to a very high degree. It should be emphasised that most of these calculations were done using a slide rule and with the assistance of 'computers' – women mathematicians who were found to have a particular flair for complex calculations using paper and pencil – and mechanical calculators that often were slower and less precise than the human equivalent. (NASA)

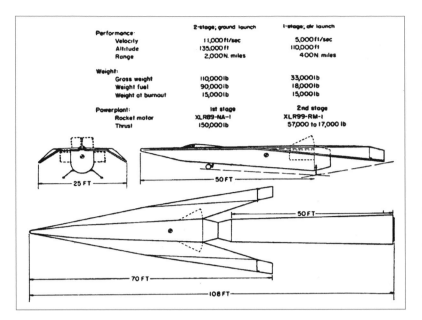

Performance:	2-stage; ground launch	1-stage; air launch
Velocity	11,000 ft/sec	5,000 ft/sec
Altitude	135,000 ft	110,000 ft
Range	2,000 N. miles	400 N. miles
Weight:		
Gross weight	110,000 lb	33,000 lb
Weight fuel	90,000 lb	18,000 lb
Weight at burnout	15,000 lb	15,000 lb
Powerplant:	1st stage	2nd stage
Rocket motor	XLR89-NA-1	XLR99-RM-1
Thrust	150,000 lb	57,000 to 17,000 lb

ABOVE Development of hypersonic air vehicles was considered just around the corner in the late 1950s and in 1957 NACA worked up the concept for a Mach 10 demonstrator launched either from the ground or on a rocket. It was through a convergence of research types such as the X-15 and of demonstrators such as this, and with the early development stages of Dyna-Soar, that the progression from winged flight within the atmosphere to winged flight into space was believed to flow. *(NACA)*

BELOW The need for a ballistic capsule in an age of emerging spaceplanes and boost-glide vehicles was pursued so that physiologists and psychologists could study the impact of space flight on the human body. It was an essential precursor to the operational use of winged vehicles in space. Here a test subject simulates the piloting actions of a spaceman while medical technicians monitor his life signs. *(NASA)*

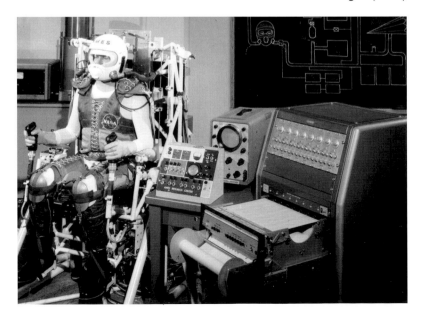

and defined by the Air Force as a subset of Dyna-Soar. Why, now, should it go to a neophyte created out of an advisory committee?

On 29 July, the day Eisenhower signed the Space Act, ARPA boss Roy Johnson met with NACA's Hugh Dryden and Secretary of Defense Neil McElroy to thrash out a workable future, but the meeting adjourned with no agreement having been reached. The final decision would come from the White House and the verdict was delivered on 18 August, when President Eisenhower decided that MISS and all future manned projects should go to the emerging NASA when it metamorphosed from NACA on 1 October.

The reason why he did this can be attributed to three identifiable imperatives in the President's thinking: Eisenhower wanted a very clear demarcation line between ostensibly civilian and obviously military programmes and he saw MISS as a project to gather scientific and physiological data; he wanted to give the new space agency a confident start by giving it a programme already defined and in need of articulation; and, fiscally conservative, he probably balked at the direct linear association with the three main Air Force mission expansion plans leading all the way to bases on the Moon, and was disinclined to start a drain on treasury coffers by giving such ambitious goals inertia.

The transition of the first US manned space project from the Air Force to NASA, the civilian 'space' agency, was a gradual process made not a little painful as the Air Force saw its grip on the future slowly slipping away. As a direct result of the furore over the Sputnik flights the country's political leadership wanted action, but the President had been reluctant to get dragooned into a race with the Russians. Eisenhower knew, as the public and almost all of government certainly did not, that America already had a robust, albeit clandestine, space programme serving the immediate future requirements of the intelligence agencies with spy satellites.

Beyond that, the President could see only limited value in human space flight per se. While supporting a measured and fiscally conservative programme for unmanned satellites and space probes, the administration had been supportive of Project 7969 only while it underpinned essential research for the Dyna-Soar boost-glide system. While NASA would interpret

RIGHT The essential design characteristics of
what would become the Mercury spacecraft
emerged from origins within the Air Force ballistic
capsule support programme, with a lattice tower
support for an escape rocket to lift the astronaut
to safety in the event of a problem with the
launch rocket. *(NACA)*

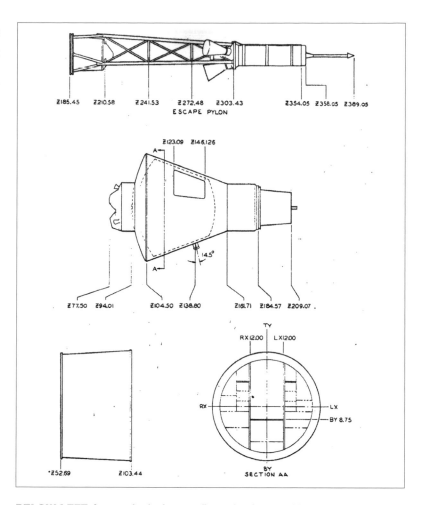

their mandate to regard Mercury as the launch
pad for human space travel, Eisenhower saw
Mercury as being on the critical path to Dyna-
Soar, and nothing more.

It was for this reason that the President was
reluctant to see manned space flight beyond
Mercury as an essential activity requiring
government support, and because of that he
could not support NASA's desire to build on
the one-man flights and engage in manned
space flight for its own sake. When Eisenhower
left office in January 1961 Dyna-Soar was in
funded development and Mercury was about
to fly its first astronaut into space. It would fall
to his successor to first downgrade Dyna-Soar
to experimental status (designated X-20) in late
1962 and a year later to cancel it outright.

But over time it would be seen that Dyna-
Soar's greatest legacy was enshrined within
NASA's Mercury programme, bequeathed
by the Air Force; transformed from being an
experimental adjunct to spaceplanes to become
the foundation stone of the human space flight
programme for which NASA became famous.

BELOW LEFT A capsule design configuration is tested in the Lewis
Laboratory's 18in x 18in wind tunnel. *(NASA)*

BELOW Langley Aeronautical Laboratory tests the configuration in the
300mph 7ft x 10ft tunnel. *(NASA)*

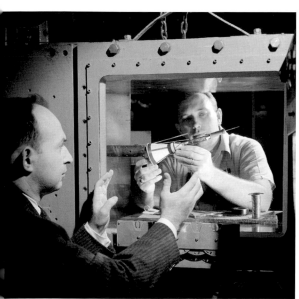

Chapter Two

Development (1958–61)

The now familiar shape of the Mercury spacecraft was defined by two discoveries: that a blunt body fabricated from known materials would provide the optimum shape for surviving the heat of re-entry; and that it would provide maximum deceleration by optimising lift over drag.

OPPOSITE The Mercury configuration is tested at the Langley Research Center. *(NASA)*

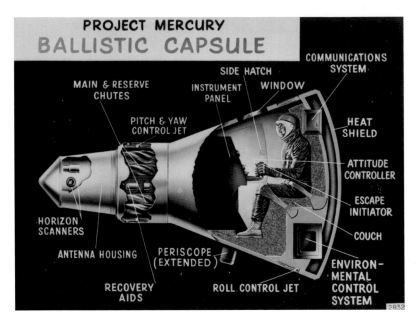

PROJECT MERCURY
BALLISTIC CAPSULE

MAIN & RESERVE CHUTES

SIDE HATCH

INSTRUMENT PANEL

WINDOW

COMMUNICATIONS SYSTEM

PITCH & YAW CONTROL JET

HEAT SHIELD

ATTITUDE CONTROLLER

ESCAPE INITIATOR

HORIZON SCANNERS

COUCH

ANTENNA HOUSING

PERISCOPE (EXTENDED)

RECOVERY AIDS

ROLL CONTROL JET

ENVIRON-MENTAL CONTROL SYSTEM

ABOVE The evolution of the first American manned spacecraft from the Air Force MISS concept to the design of NASA's Mercury spacecraft began in 1958 and progressed through to early 1959 when the definitive configuration was agreed. *(NASA)*

In its role as a research organ of the government, NACA applied its facilities, laboratories and test sites to a wide variety of scientific and engineering problems outside the general area of flight and flying. Among these were finding solutions to problems involving very high-speed flight. Just as NACA was thinking about hypersonic flight research, bridging the zone between aerodynamics and thermodynamics, some researchers at the Ames Aeronautical Laboratory were studying the problems encountered by high-speed aircraft such as the forthcoming X-15.

Using specially adapted wind tunnels at Ames, NACA engineers sought solutions to aerodynamic heating on a body travelling through the atmosphere at great speed. Undertaken as part of essential research into what would popularly, and incorrectly, be termed the 'heat barrier', this work attracted the interest of engineers seeking solutions to the problem of getting re-entry warheads back through the atmosphere.

Rockets with even modest range had an apogee on their trajectory outside the atmosphere but at velocities where kinetic energy on re-entry did not produce significantly high temperatures. The V-2 is one example: rockets of this type were arching over to a peak altitude at the fringe of space by 1943. But rockets designed to throw a warhead several thousand miles would fully exit the atmosphere and reach altitudes as high as 900

miles (1,450km), achieving velocities as high as 12,000mph (19,300kph). The stagnation temperature in the shock wave of such a body would reach 6,650°C (12,000°F), about ten times the temperature that would be experienced by the hypersonic X-15.

Early rockets like the V-2 had warheads integral with the forward section of the projectile, but instability caused by the significant change in its centre of mass as its propellant was depleted caused tumbling. This frequently resulted in the destruction of the entire vehicle as it re-entered the denser layers of the atmosphere, warhead included. Warheads that could be made to separate from the main body of the rocket after burnout were the solution, leaving the inert tankage and rocket motor to be destroyed during re-entry. But this still left the problem of how to make a warhead capable of surviving the extremes of heat caused by atmospheric friction.

The breakthrough came as a result of the work of Harry Julian Allen, who held a degree in aeronautical engineering from the Guggenheim Aeronautical Laboratory, and moved from Langley to Ames when it opened fully in 1941 as a dedicated wind tunnel research facility. Allen was largely responsible for the construction of a supersonic free-flight wind tunnel in 1949 with a test section 18ft (5.5m) long, 24in (61cm) high and 12in (30.5cm) wide, which operated by forcing air up through the slot at Mach 3 and firing variously shaped projectiles from the other end at 5,450mph (8,770kph). The characteristics of the resulting airflow were observed on a series of photographs taken by Schlieren cameras positioned at various locations.

Industry studies into the problem employed early computers to predict the optimum configuration for an entry vehicle – be it warhead or spacecraft – and came up with a sharply pointed object. However, when models of this configuration were subjected to tests in the Ames supersonic wind tunnel they showed that thermal transfer to the body of the model would cause it to melt, even with the most heat-resistant materials then available. It did seem, for a while, that the 'thermal barrier' was indeed just that. But several engineers, Allen among them, regarded that as a product of

ignorance rather than physics, as the 'sound barrier' had been a few years earlier.

Allen did some manual calculations and realised that the computers had been too limited in their run and that if left to continue producing calculations the solution would have become evident, a truth unravelled by Allen using slide rule, pencil and paper! Allen was frustrated by the early results which showed that half the frictional energy produced ended up in the body of the sharp-nosed projectile, which to date had been the shape of dummy warheads tested in smaller rockets.

Allen discovered that the secret lay in deflecting the heat back out into the airstream, and he found a way to do that. By presenting a blunt, slightly convex shape to the airflow the shock wave that would form would separate from the physical structure, with a boundary layer between the shock front and the structural body of the re-entry vehicle. In this way a large proportion of the heat would be carried away by the shock wave itself, comfortably detached from the body, and not conducted into the structural material.

This was contrary to traditional teaching in aerodynamics and turned the problem on its head to obtain the solution. The findings were convincingly presented to a select group of engineers from the rocket and missile industry in September 1952. Over the next several months a technical paper was worked up by Allen and Alfred J. Eggers, also at Ames. Completed on 28 April 1953, it was distributed as a secret document to engineers working on ballistic warhead design for the coming generation of long-range missiles.

As debate evolved around the use of re-entry vehicles for returning humans from space, especially the Air Force Man-In-Space-Soonest proposal, issues surrounding the type of materials flared up. Despite the Allen/Eggers

BELOW Wind tunnel tests were crucial to determining the degree of oscillation as the capsule descended through the atmosphere. Subtle changes to the shape were tested, particularly the lip of the heat shield/forebody interface. *(NASA)*

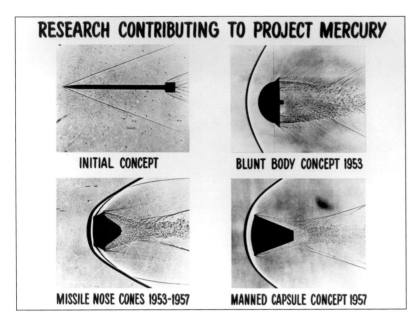

RESEARCH CONTRIBUTING TO PROJECT MERCURY

INITIAL CONCEPT

BLUNT BODY CONCEPT 1953

MISSILE NOSE CONES 1953-1957

MANNED CAPSULE CONCEPT 1957

LEFT The evolution of the blunt-body concept from a theoretical idea in 1953 to its application on manned spacecraft design. The initial concept imposed thermal conduction on the tip of the missile, whereas a blunt body relieved the structure of direct thermal conduction by detaching the shock wave. From that to the design of the Mercury spacecraft was a logical step. (NASA)

BELOW The overall configuration of the spacecraft was dictated by the physics of thermodynamics but the precise design incorporated essential safety features such as the launch escape tower, an egress hatch and retrograde rocket package. (NASA)

blunt-body concept, temperatures would still be very high. At the start of re-entry, a blunt body would create a shock wave with a temperature of 9,500°F (5,260°C), while the blunt face of what constituted a heat shield would be at a temperature of 3,000°F (1,650°C).

There were just two principles governing the category of materials that could safely protect warheads and incoming re-entry vehicles: heat sink and ablation. Existing materials such as aluminium and titanium were universally applied to the design of hot structures experienced by X-series NACA research aircraft, but new materials, such as Monel K and Inconel-X, were already being investigated as thermal protection for hot structures using the heat-sink method. Ablative materials would char away and take

with them the heat conducted across the boundary layer.

NACA looked at various ablative materials, and Teflon, nylon and fibre glass composites were tested as well as heat-sink materials such as copper, beryllium and tungsten. But the Air Force conducted its own tests using the X-17, a specially built research rocket from Lockheed Aircraft Corporation, using warhead materials from General Electric for heat-sink concepts and from Avco Corporation for ablatives.

Throughout 1955–56 NACA-Langley exposed a range of materials to temperatures of up to 4,100°F (2,260°C) in the plume of an acid-ammonia rocket jet at a gas stream velocity of 2,133m/sec (7,000ft/sec). High-temperature electric arc jets at Langley and at the Lewis Flight Propulsion Laboratory provided energy for compressed air and for increasing the pressure and temperature of the air, producing a steam gas temperature of 12,000°F (6,650°C).

Research into hypersonic flight and into the possibility of returning men from space was fundamental to NACA's mandate. Members of the Pilotless Aircraft Research Division at Langley preferred to pursue their research not in wind tunnels but with small and medium-size rockets, where they could fly a variety of different configurations in the search for an optimum shape for bodies returning from space and for conducting a controlled descent to a predetermined landing site. This was fundamental to their purpose.

Maxime A. Faget and Paul Purser led a small team working for Robert R. 'Bob' Gilruth at PARD, and together they produced small-scale models of Allen's blunt-body shapes to see how they performed in flight. Robert O. Piland assembled the first multistage rocket capable of reaching Mach 10, a thin, relatively small research projectile. Faget and his colleagues

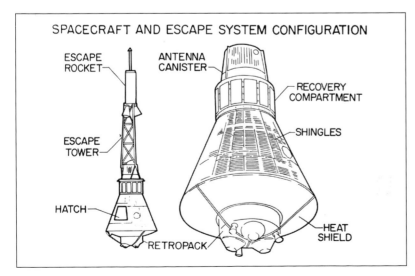

SPACECRAFT AND ESCAPE SYSTEM CONFIGURATION

ESCAPE ROCKET

ANTENNA CANISTER

RECOVERY COMPARTMENT

ESCAPE TOWER

SHINGLES

HATCH

RETROPACK

HEAT SHIELD

worked out of Wallops Island test facility, a small launch site situated on the shore of a peninsula off the coast of Virginia.

A part of the Virginia-barrier islands, Wallops was managed by NACA-Langley and it was from this lonely site that sounding rockets designed to study the upper atmosphere were frequently launched. It was an ideal place to conduct practical flight tests and to experiment with heat loads and heat transfer and to work on the development of heat-sink materials that could not only protect warheads but also pilots returning from space.

The PARD group would fiercely defend this means of acquiring data, claiming that theoreticians at desks could only visualise situations and extrapolate from known data in areas where conclusions were unproven by practical tests. Greater progress could be made, they said, through the build, fly and test ethic of the experimenter. Backed with research results, they would make enormous strides in understanding the optimum shape of bodies returning from orbit, aspects of flight and flying for which the Air Force itself sought quick results.

Countering this PARD-centric view of experimentation, Eggers developed a new instrument at NACA-Ames that would simulate the aerothermal environment at an altitude of 100,000ft (30,480m), a height at which he believed most of the heating would be encountered. The new supersonic tunnel had a maximum diameter of 50.8in (20cm) and a length of 20ft (6.1m) in a scaled representation of the desired altitude. The tests worked when a scale model was launched to a settling chamber by an upstream hyper-velocity gun, the model passing through increasingly denser air as a result and thus simulating re-entry from space. This had the advantage of simulating

ABOVE Despite the logic of a truncated cone, several studies continued to carry out research on a lift-body configuration that would have significant aerodynamic manoeuvrability in the transonic and subsonic regime. This M-1 design was tested in a high-enthalpy airstream at the Ames Research Center. This work would lead eventually to the Lifting Body series of NASA research vehicles and provide additional data bolted on to X-15 research results that would lead to the Shuttle concept in the late 1960s. *(NASA)*

both movement and heat. Using a model only 0.36in (0.9cm) in diameter weighing 0.05lb (22.7g), this 'atmospheric entry simulator', as Eggers called it, could simulate a 5,000lb (2,268kg) object with a range of 4,000 miles (6,440km) slicing into the atmosphere. The test rig went into operation during 1955 and a larger version was started three years later.

RIGHT Before final configuration specifications were written essential characteristics were decided by the Space Task Group out of Langley Research Center, renamed from Langley Aeronautical Laboratory on the inception of NASA. Bidders were not required to adhere directly to the recommended characteristics, but these were projected as the preferred baseline. *(NASA)*

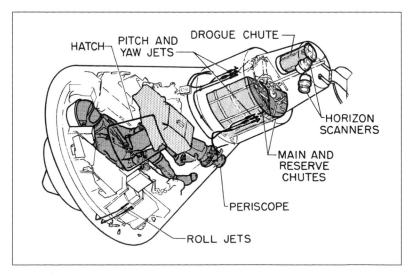

HATCH

PITCH AND
YAW JETS

DROGUE CHUTE

HORIZON
SCANNERS

MAIN AND
RESERVE
CHUTES

PERISCOPE

ROLL JETS

Shaping for the future

Elsewhere at NACA work was being conducted on the control systems for a manned vehicle outside the Earth's atmosphere, where aerodynamic surfaces such as ailerons, elevators and rudders were useless in the vacuum of space. Much of this work was funded in support of the development of the X-15 rocket research aircraft, then being designed to carry its occupant to altitudes exceeding 66.3 miles (106.7km), where there was insufficient atmospheric pressure to provide attitude orientation and stabilisation. Reaction control thrusters were the sole means of accurately maintaining a stable attitude above about 19 miles (30km), an attribute essential when returning from space to stabilise the vehicle as it encountered the atmosphere. Studies of blunt body shapes emphasised the necessity for accurately maintaining specific attitude and pointing angles as the object re-entered the atmosphere, to prevent it tumbling and burning up.

As early as 1954 the PARD team at NACA-Ames had prepared an advisory paper defining the list of possible configurations for re-entry vehicles for advanced applications. They evaluated relative advantages and disadvantages of different configurations and various ways of returning them through the atmosphere, determining the effect of weight, velocity and angle of entry, examining ballistic vehicles, blunt body shapes as high drag devices, atmospheric skip designs and glide vehicles. Both skip and glide concepts possessed some degree of lift but were optimised with blunt nose shapes and the high angle of attack (pitch-up) required during entry interface, the point where the incoming vehicle began to feel deceleration due to atmospheric friction inducing drag.

Assisted by Stanford E. Niece, Allen and Eggers discovered that the skip and glide

ABOVE Development tasks included boilerplates used for weights and balance tests, mass distribution studies (essential to determining where to pack the essential equipment), vibration analysis and pendulum characteristics induced on separation in the event of an in-flight abort of the launch vehicle. *(NASA)*

LEFT Boilerplate capsules duplicating the mass and balance characteristics of a flight-rated spacecraft were also prepared for launch on rockets, including Little Joe on an abort simulation such as that to which this one has been assigned. *(NASA)*

concepts would provide the greatest range of design options but that each would have serious disadvantages. The skip concept would call for the vehicle to rise in and out of the atmosphere, 'skipping' like a flat stone on the surface of water. But it would require significant boost to circumnavigate the Earth and make increasingly steeper penetrations into the atmosphere, generating high heat load, with successively higher heat rates presenting very high aerodynamic heat loads.

Heat load and heat rate would be the two irreconcilable parameters required of materials selected for thermal protection during re-entry and would come to characterise the selection of incoming trajectories both for Earth-orbit vehicles and those returning from deep space. Conversely, the glide concept would ensure the pilot had authority over the vehicle all the way down to a controlled landing, with acceptable heat loads countered by thermodynamic loads that could significantly increase the internal temperature of the vehicle.

While it was almost impossible to control the ballistic-shaped capsule during the descent phase it represented the simplest configuration of all three, possessing relatively benign g-loads where deceleration during re-entry could be kept well below 10g at most and long periods at less than 3g. The limitations of evolutionary growth with a ballistic capsule were also noted and the dead-end approach of a simple ballistic capsule favoured the more difficult technical approaches of a skip or a glide vehicle, albeit with their penalties.

Despite the difficulties, the chosen configuration was a delta-shaped boost-glide vehicle very similar in plan form to the Dyna-Soar, which would return with a high angle of attack to dissipate thermal energy from kinetic heating, pitching down in the lower levels of the atmosphere to increase the lift-over-drag (L/D) ratio. The glider would have rounded delta-wing leading edges to inhibit airflow attachment of the shock wave to the structure, to minimise thermal loading on the wing itself.

Nevertheless, Eggers in particular was worried about the high forces of deceleration of the glider concept and the excessive thermal loads on the structure of this configuration. So he came up with a compromise, a single vehicle shaped approximately like a flat iron but with a rounded top and a slightly bulbous under-surface and blunt nose section. Designated the M-1, it was the birth of the 'lifting body' research vehicle which NASA would test in the 1960s and from which would evolve configurations prior to development of the Space Shuttle.

During 1956–58 this work fed directly into Air Force plans for the Dyna-Soar boost-glide system and to the Manned Ballistic Rocket Research System expected to provide physiological data on pilots. It was for this reason, and from the technology research results on shapes and materials, that the Air Force manned ballistic capsule, designated Task 27544, took the form of a truncated cone.

Throughout 1958, as it became clear that the old NACA would be taking over several space-related projects from other government bodies such as the Air Force and the Army, several key individuals at the Pilotless Aircraft Research Division seized control of the opportunity and married their original research work on ballistic re-entry vehicles to the accelerating interest in getting a man in space as quickly as possible. The one key individual who made that connection was Maxime Faget, the most influential engineer on the design of what became the Mercury spacecraft.

Design solutions

The origin of Project Mercury is hard to define in terms of a single date. It had been formed as an idea out of the Air Force's requirement for a simple way of getting basic data about humans operating in space prior to the development of a space weapon system. But as a project undertaken by the nation's civilian aerospace research organisation, it was probably on 18 March 1958 when Faget, along with Benjamine Garland and James J. Buglia, published a report with the title 'Preliminary Studies of Manned Satellites, Wingless Configuration, Non-Lifting'.

This coincided with a conference held at Ames Aeronautical Laboratory on 18–20 March examining several proposals from different NACA centres on how to get pilots into space and in what sort of vehicle. Several proposals focused on lifting bodies and on glide vehicles, since this was where most of the

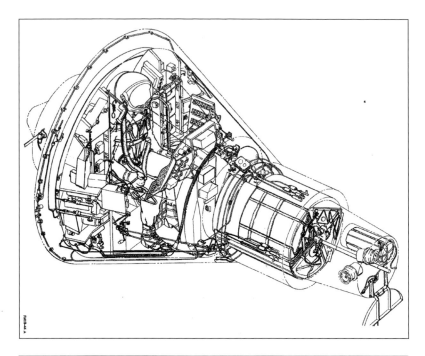

original research had been focused over the preceding decade. The conference focused on a programme plan laid out by NACA for taking over this research from other agencies and mirrored the political moves made during the year to create the new space agency.

At Langley Aeronautical Laboratory, a study group was set up during March 1958 which concluded that a ballistic capsule launched by an adapted ICBM – the only two available for consideration being Atlas and Titan – was the most logical way to go as America's first manned spacecraft. In this, the project metamorphosed from a support for Dyna-Soar under the MISS programme to the first step on the ladder to the Moon. Similarly, NACA's Bob Gilruth, Clotaire Wood and Hartley A. Soule sent a document to the ARDC advising of the preferred concept and recommending that the thermal shield should be of a heat-sink design.

Several key design trends began to emerge around this time, including the adoption of a contour couch with its occupant lying with his back to the direction of flight during re-entry, so that forces of deceleration would be felt from back to chest rather than head to toe. In this way the g-loads would be uniform across the whole body, with head, torso and limbs supported across their full surface area. While unconventional in principle, the idea

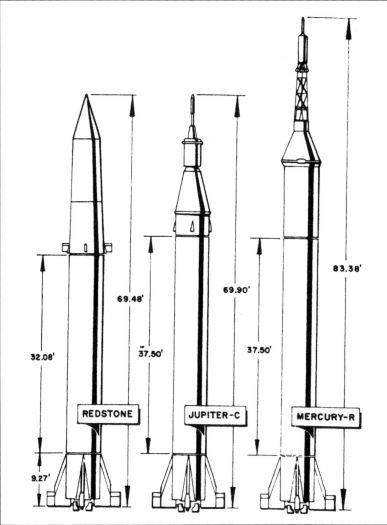

REDSTONE — 32.08' — 69.48' — 9.27'

JUPITER-C — 37.50' — 69.90'

MERCURY-R — 37.50' — 83.38'

had traotion. On 00 July test subject Carter C. Collins withstood a 20g load in a contour couch during acceleration tests on the centrifuge at Johnsville, Pennsylvania.

The conference of 18–20 March resolved contentious issues surrounding the choice of a ballistic capsule, yet the very nature of the concept had specific advantages. A ballistic trajectory implied minimal requirements on automated attitude and guidance and control with the added advantage of reduced weight, which was already a critical issue. The maximum orbital lift capability of an Atlas ICBM was approximately 4,000lb (1,800kg) and the engineering pressures to cut this weight to a minimum were crucial to its viability. Moreover, retrograde (retro) rockets would be sufficient in themselves to bring the spacecraft out of orbit, the deceleration of atmospheric braking being sufficient to reduce the descent rate to a level within the capture velocity of parachutes.

By May there was a general consensus among the PARD team that the capsule would have a height of around 3.35m (11ft) with a diameter of 2.13m (7ft), sufficient for a single occupant on a flight lasting up to a day in orbit. Strapped tight inside his contour couch, the pilot would face forward during ascent and upon entering orbit the capsule would be turned around so that its occupant would face backwards until the time came to fire braking rockets to slow the speed and cause the tiny spacecraft to fall back on an arching path toward the atmosphere.

As refinements were added to the overall tests for slightly varying configurations, problems appeared with the initial design, which had favoured the general shape of the Mk 2 re-entry warhead with a recessed body inside the lip of a base heat shield, which it was thought would prevent turbulent air flowing back around the afterbody and imposing high heat loads on the conical sides. Tests with this design in the free-spinning wind tunnel at Langley showed poor stability at subsonic speeds and the lip was eliminated.

The final design had a base heat shield with a maximum diameter of 6.67ft (2.03m) and a radius of curvature of 120in (305cm) possessing a ratio of 1:5 between the radius of the curve and the heat shield. This replaced

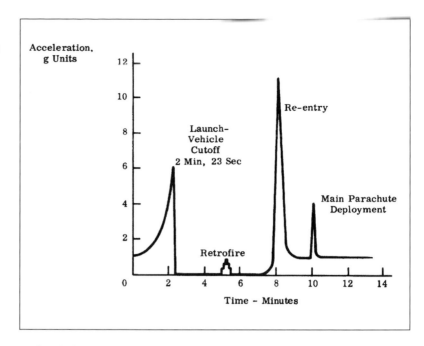

earlier designs with a flat heat shield, which was found to trap excessive amounts of heat, and a rounded surface, which increased the transfer of heat. The conical afterbody supported a cylindrical drum mounted on top of the frustrum in which were situated the recovery parachutes. The design had been finalised by William F. Stoney Jr from PARD and Alvin Seiff and Thomas N. Canning from the Vehicle Environmental Division at NACA-Ames.

A new day

A day after NASA opened for business on 1 October 1958 personnel from the renamed Langley Research Center visited the Army Ballistic Missile Agency to discuss the possibility of procuring ten Redstone and three Jupiter (not to be confused with Jupiter-C) rockets to conduct test flights of boilerplates and test capsules with primates and pilots, as well as qualification of the launch escape and parachute recovery systems. While the Atlas ICBM would serve as the launch vehicle for orbital flights of the capsule, they were expensive and unnecessary for ballistic shots required during the development cycle.

Back at the beginning of the year, Paul E. Purser and Max Faget conceived a purpose-built solid propellant rocket combination that could be used for evaluating the capsule under the dynamic loads and flight conditions a

ABOVE The accelerations imposed by the Redstone flight profile were the highest encountered in the Mercury programme, with the ballistic trajectory imposing more than 11g on the astronaut during re-entry. *(NASA)*

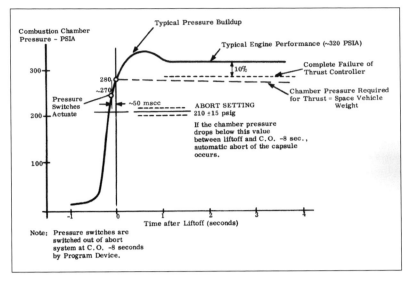

Note: Pressure switches are
switched out of abort
system at C.O. -8 seconds
by Program Device.

ABOVE Man-rating of the Redstone included a set of sensors to trigger an abort in the event of any malfunction, most critical of which was a loss of chamber pressure which could result in low thrust and the collapse of the rocket back on to its launch pad. This chart displays the critical values that would automatically trigger an abort. *(NASA)*

BELOW In a suborbital ballistic test of the Redstone rocket, the spacecraft would follow an arching trajectory carrying it high into space and back to a recovery virtually identical to the terminal stage returning from space, albeit at slower speed and with considerably less thermal load. The spent booster case would hit the water farther downrange than the spacecraft. *(NASA)*

manned capsule would experience. It was to be called Little Joe and it would be put to good use in the months to come, doing much of the test qualification work, an example of how the PARD team liked to do things through practical experimentation and flight evaluation rather than theoretical extrapolation.

On 8 October an informal decision was made to set up a Space Task Group (STG) comprising the core PARD engineers who had pioneered the manned capsule concept and worked so hard for much of the year to define its specification and optimum configuration. It would be headed by Bob Gilruth, supported by Charles J. Donlan as his assistant, with Max Faget closely involved in detailed definition of the spacecraft.

Gilruth and Donlan were key to getting the STG organised and formed a small group to act as executive council while others worked up the technical plan for developing the spacecraft itself and organising an operating plan for flight tests and manned launches. The STG itself formed up with 33 people from Langley under Gilruth and Donlan of whom 14 of the 25 engineers came from PARD. The official date of transfer was 5 November. Technically responsible to NASA HQ, the STG was in fact

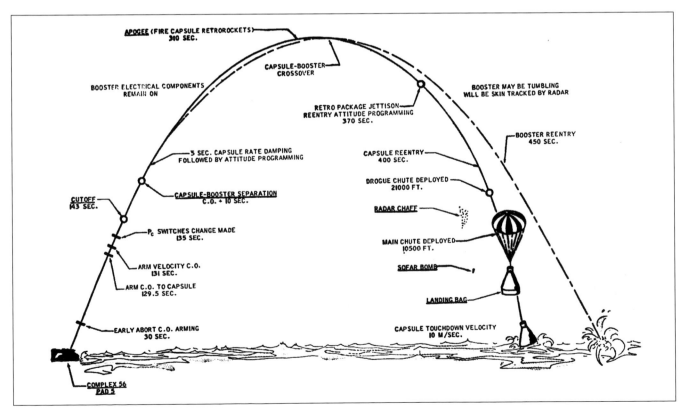

run from Langley Research Center and soon began to acquire additional staff, particularly from the now renamed Lewis Research Center.

The contract would be awarded to the winner in a series of proposals solicited through bids covering a specification defined by the STG. A Technical Assessment team was formed under Donlan to manage 11 separate categories, a separate group marking up bids in each. These areas consisted of systems integration; load, structure and heat shield; launch escape system; retrograde and landing system; attitude control; environmental control; pilot support and restraint system; displays and navigational aids; communications; instrumentation sensors, recorders and telemetry equipment; and power supplies. Each category could get a maximum of five points on a scale from unacceptable to excellent with a summed average to produce an overall figure.

The formal process for selection of a contractor to build the Mercury spacecraft began on 20 October 1958 and preliminary specification and requirements were mailed to the more than 40 prospective bidders on 23 October, from which group 38 attended a special conference at Langley Field on 7 November. Seven days later a refined requirements document consisting of 50 pages was mailed to the 19 firms who remained interested in bidding. Three days before it was sent eight of these firms had dropped out. The 11 remaining had their bids in by the 11 December deadline but it would be a month before the winner was selected.

Eight companies had shown serious interest in competing: Avco, Convair Astronautics, Lockheed, Martin, McDonnell, North American Aviation, Northrop and Republic. Three with less committed resolve were Douglas, Grumman and Chance-Vought. Some were so heavily committed elsewhere that they would be unlikely winners: Bell with Dyna-Soar; Boeing with the Minuteman ICBM and support work on Dyna-Soar; and, being a parent company for Pratt & Whitney, Hamilton Standard and Sikorsky, United Aircraft had a very full workload already. Meanwhile there were rockets to choose, for tests and for operational manned flights.

The decision on boosters was shaped largely by availability and cost. Little Joe had

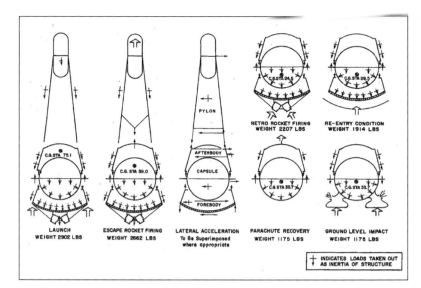

already emerged from in-house work at the PARD group in NACA, but NASA would have to procure several rockets that came directly from the Department of Defense. Only the Army, the Navy and the Air Force had vested interests in rockets and missiles capable of supporting a manned space programme. The first order was placed on 24 November 1958 for a single Atlas C with the intention of using this as a test vehicle for the manned spacecraft. The order was shifted later to an Atlas D, the first of several Atlas rockets eventually ordered.

This first Atlas would support the Little Joe flight-test programme by providing the energy required to drive the capsule back down through the atmosphere from an orbital velocity of 17,000mph (27,350kph). Little Joe and Redstone could propel test models of the capsule to orbital altitude but only to 4,100mph (6,600kph), much less velocity than would be experienced during manned orbital flights. It was therefore unable to subject the heat shield to a realistic environment that it would experience during manned operations. And there were still many unanswered questions about the behaviour of materials subject to these conditions, which was why there was interest in flying Jupiter test shots capable of boosting the capsule to 10,900mph (17,500kph).

Ballistic missiles were designed to send their warheads at steep angles through the atmosphere, minimising the loss of speed due to atmospheric friction and in so doing drastically reducing the effects of strong winds on blowing it off course before reaching its target, but

ABOVE When NASA went out to industry for bids on building the Mercury spacecraft it invited suggestions on alternate configurations and concepts. Here Avco Corporation – which specialised in heat shield materials – adopted a spherical pressure cabin encased within a truncated capsule, with a tractor escape system pulling the spacecraft free in the event of an abort. *(Avco)*

experiencing high heat rates. The manned capsule was designed to do the opposite, to slice into the atmosphere at a very shallow angle, using the atmosphere to slow it down but incurring much greater heat loads before deploying parachutes to slow its descent for splashdown.

Some of the more critical questions raised by these concerns included the behaviour of the convex ablative heat shield and the degree to which heat from re-entry would flow back, carried on turbulent air across the conical after body. The STG design adopted shingles to minimise the effect of these turbulent flow patterns, which it was felt could become attached to the body of the capsule, but nobody knew for sure how the spacecraft would react during its long descent through the atmosphere.

Early studies based on theoretical calculations appeared to indicate that the heat rate on these shallow re-entry trajectories would be an order of magnitude lower than those already being tested on ballistic missile re-entry vehicles.

At the end of 1958, when the STG was getting ready to send the specification to industry for bids, NASA was still debating the ablation versus heat-sink method for thermal protection during re-entry. In fact the results so far were contradictory. The Army used ablative materials such as graphite, Teflon and nylon to protect the nose cones of its Redstone and Jupiter missiles, but the Air Force had favoured copper as a heat-sink material for Atlas while the Navy found that beryllium worked well for tests with re-entry vehicles for their Polaris submarine-launched missiles. The Faget design group favoured the heat-sink concept, with an added advantage that the plasma sheath that enveloped the re-entry vehicle seemed to make the blocking of communication with the capsule less severe. The decision as to which was preferable was pending and the debate would drag on, although the original specification given to bidders on the Mercury contract specified ablation.

Bids and contracts

The prime contract was decided by a group of men who met during the first week of 1959. Led by Carl Schreiber at NASA headquarters in Washington they judged the four companies still in the running, but in the final analysis it was Abe Silverstein and his Source Selection Board (SSB) who would decide. A long-term director of NACA's Lewis Flight Propulsion Laboratory, on the formation of NASA Abe Silverstein had been appointed Director of Space Flight Development and had been closely involved in the formation of the agency out of the old NACA.

The two competing bids were from McDonnell and Grumman, the latter a veteran builder of strong US Navy amphibious flying boats and carrier-based combat aircraft, having produced the war-winning Avenger and Hellcat and very recently won contracts for the E-2 Hawkeye airborne early-warning aircraft and the A-6 Intruder for penetration strikes. Formed before America entered the Second World War, McDonnell was

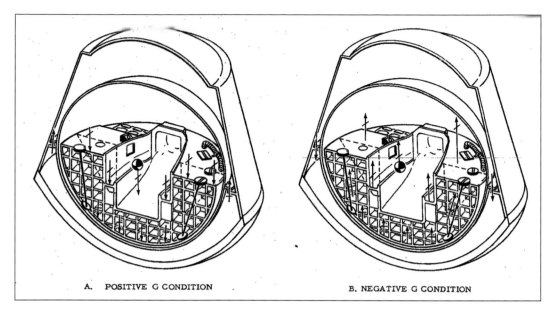

A. POSITIVE G CONDITION   B. NEGATIVE G CONDITION

LEFT Grumman had a novel solution to the g-force problem that would be experienced on re-entry, incorporating a spherical pressure cabin that could pivot and rotate to place the occupant in the chest-to-back vector for accelerations. *(Grumman)*

in truth a child of the nascent jet age and had made a big reputation with the FH-1 Phantom, the F2H Banshee and the F3H Demon, all shipboard fighters of impressive performance. Now it had the F4H-1 Phantom II, which by the end of 1958 had been selected by the Navy as its first supersonic fighter (Mach 2+) and tactical strike aircraft.

The SSB decided in favour of McDonnell due to the extensive contract work already under way at Grumman, which it was felt was already stretched tight on manpower, which might compromise a quick start on the Mercury spacecraft. Accordingly, McDonnell were selected to build the manned capsule and were so informed on 12 January. In terms of aircraft procurement it was a relatively small amount of money, the estimated development price of $18.3 million ($152.4 million in 2016 money) incorporating a fee of $1.15 million ($9.5 million today). At the end of the programme the total bill for Mercury for all costs to all contractors and reimbursement to the Department of

RIGHT Lockheed's proposal was ultra-conservative and followed the specified configuration most closely, although it did incorporate an 'astro' window in the hatch as well as the mandatory periscope. It did, however, place the escape tower jettison motor on the side of the stand-off truss and had an erectable high-frequency antenna which it proposed to eject prior to re-entry. *(Lockheed)*

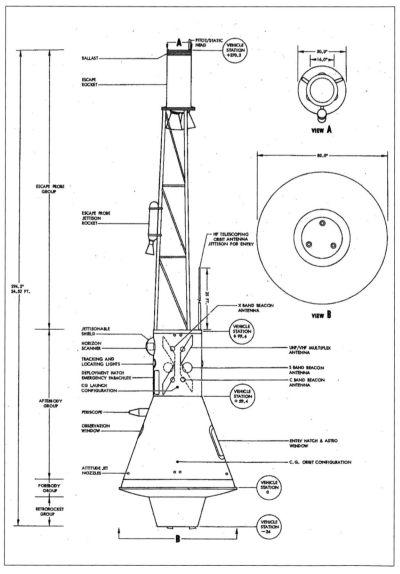

RIGHT Lockheed's periscope incorporated a fixed support mounting penetrating both spherical pressure cabin and outer skin. The optics were particularly well designed and served to stimulate further in-house development at NASA, working with the winning bidder on incorporating some valued features. *(Lockheed)*

RIGHT Martin came up with a novel weight-saving system: a conformal escape motor configuration whereby the solid propellant rockets were placed within a shroud encapsulating the forward end of the spacecraft. If not employed for an abort they would be utilised as retro-rockets to de-orbit the spacecraft. This concept would be reintroduced for the SpaceX Dragon 2 crew capsule scheduled to carry astronauts to the International Space Station from 2018/19. This design also adopted a shaped pressure cabin rather than a spherical capsule. *(Martin)*

to 20 – each individually tailored to specific requirements. McDonnell began with an outline specification but then began to define its 15 subsystems to produce the blueprints and control drawings that it would submit to the STG. McDonnell conducted a Mock-Up Inspection Board meeting at their plant on 17–18 March 1959 where they presented an initial set of capsule specifications and a full-scale model to the STG, and based on that some 34 alterations were recommended of which 25 were immediately authorised.

Another Mock-Up meeting was held on 12–13 May where further changes were presented, and a third was held on 10–11 September where the newly-appointed astronauts participated. Their input significantly defined the way the pressurised compartment was configured and the manner in which the instruments were displayed, as well as evincing forceful ideas on the degree to which the man should be 'in the loop', controlling key features or maintaining a manual option. They were also instrumental in getting the large observation window installed, redesign of the instrument panel and in getting an explosive side hatch.

But there were differences between the way

Defense for recovery services would amount to $400 million (about $3.36 billion today).

In the initial language of the contract NASA requested 12 spacecraft – which would grow

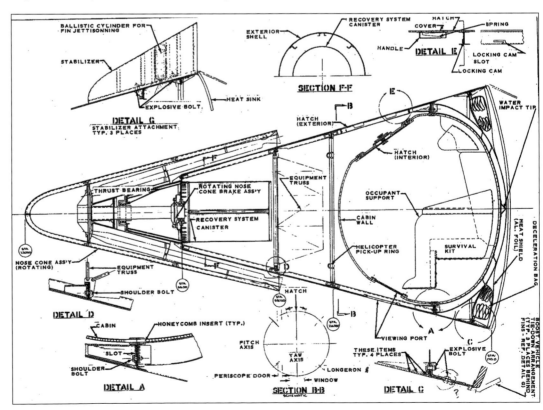

McDonnell saw the optimum design evolving and the manner in which the STG wanted to control the imposition of its own decisions – after all, it was the customer. One of these concerned the escape system. McDonnell had proposed a push concept with stabilising fins around the base, but that was rejected by NASA; the other was about the knotty issue of ablative or heat-sink shielding. The STG argued for heat-sink, and McDonnell ordered six heat-sink shields from the Brush Beryllium Company but also ordered 12 ablation shields from Cincinnati Test and Research Laboratory. The dual option was prudent because, of the two beryllium forgers approached, only Brush had managed to produce ingots of the required purity.

Engineers working the Big Joe flight for an Atlas test of the thermal protection argued for an ablation shield fabricated from phenolic resin and fibreglass, and the first half of 1959 saw a dramatic volte-face in support for the ablation concept. By mid-summer the STG was persuaded that the heat-sink concept should be nothing more than an alternate backup, and early problems with fabricating a suitable ablative shield were largely solved by July.

One overriding concern was with the high

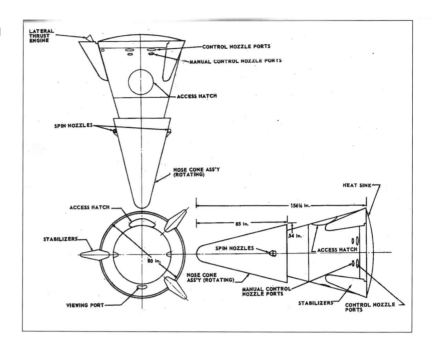

temperature retained by the beryllium shield which, while acceptable for a re-entering warhead, would be potentially dangerous for astronauts and personnel engaged on astronaut recovery activities. Faith in the notion that the heat-sink shield could be jettisoned during the parachute phase of the descent was dashed during an airborne drop-test where the jettisoned

ABOVE Martin believed the spacecraft would require stabilising fins for aerodynamic stability and placed these at the base of the conical body, surmounting the forward end with the escape/retro-rocket shroud. *(Martin)*

FAR LEFT McDonnell Aircraft included an aerodynamic fairing at the base of the spacecraft for its alternate configuration housing escape rockets, ejected on the way up to expose the retro-rocket pack. *(McDonnell)*

LEFT North American Aviation favoured a spherical pressure cabin with a circular hatch enclosed within a skin-stringer-constructed shell, supported by formers in a classic fuselage design reminiscent of its aircraft family. *(NAA)*

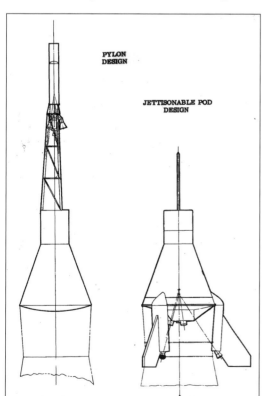

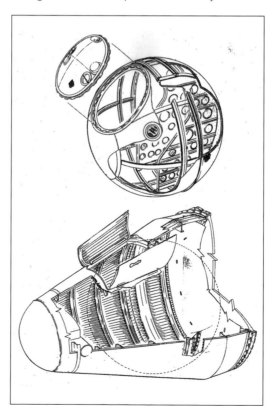

shield spiralled back and struck the capsule. But the spacecraft itself was not the only hardware in development, as adaptation of the nation's first ICBM into the launch vehicle for America's first manned spacecraft was about to begin.

If there were any doubts about the ability of an Atlas missile to depart from its design trajectory and place itself in orbit they had been removed on 18 December 1958. In a highly secret launch, Atlas 10B placed itself in orbit with a tiny package of communications equipment, a project known as SCORE (Signal Communications by Orbiting Relay Equipment). It would transmit to the world a pre-recorded Christmas message from President Eisenhower as America chipped away at Russia's Sputnik propaganda advantage. But it was a shallow boast; the flight came only three weeks after the first successful flight of an Atlas ICBM along its design range of 6,300 miles (10,135km).

In fact there were still concerns about the flight history of Atlas, although the success rate was improving. Although it would not fly for the first time until April 1959, Atlas D was the version chosen for the manned orbital flights since it was the first derivative to be capable of lifting the manned spacecraft to the required height and

velocity. Much work would be done to convert what was, in effect, a somewhat unreliable missile into a successful launch vehicle.

During the first half of 1959 key decisions were made regarding the escalating number of test flights required, as well as the number of flight systems to be procured. In March schedules were brought to draft conclusion on the required number of launches and vehicles to carry them out. There were to be five Little Joe launches, eight with Redstone, two with Jupiter rockets, ten Atlas launches and two balloon-supported thermal tests. These interleaving flights were to have occurred at various times between July 1959 and January 1961.

Designated MR-3, the first manned Mercury Redstone launch would be a ballistic 'hop' into space and back again on a mission lasting barely 15 minutes, slated for 26 April 1960. The first manned Atlas flight would be MA-7 scheduled for 1 September 1960, with progressively more protracted orbital flights up to a duration of 18 orbits, about 27 hours. It was an ambitious schedule and one that raised high expectations, within the programme and at NASA headquarters in Washington.

By May 1959 several reviews were under

BELOW Northrop came closest to the specified concept from NASA, their design being fully representative of the configuration selected by the Space Task Group. *(Northrop)*

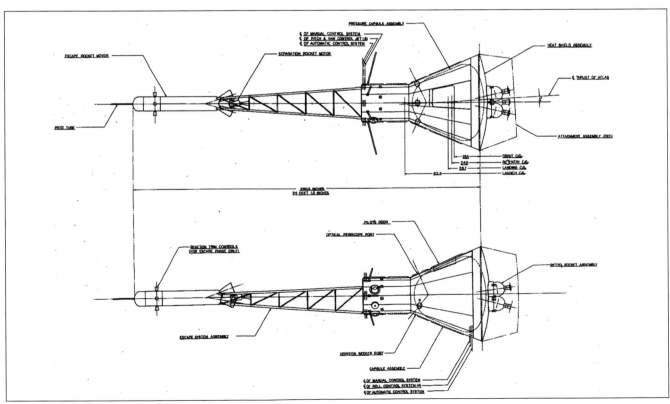

way to refine the flight test programme and to make economies wherever possible. One area of concern was the escalating price of launch vehicles, the ABMA having notified NASA of an increase of $8 million in the procurement price for Redstone and Jupiter rockets. The Jupiter was now on a cost par with Atlas and since the former could not accurately duplicate a Mercury trajectory it was decided to drop it and rely on Redstone and Atlas flights. Planned balloon flights to thermally simulate the environment at 80,000ft (24,400m) before releasing capsules to a parachute descent were also terminated. The combination of reduced costs and streamlined test schedules eased other loads on the programme through diminished complexity.

Throughout the Mercury development phase, one defining issue dogged the nascent NASA: that embracing the definition of quality control, mission assurance and safety and reliability considerations. It was one of the most difficult areas to encapsulate and produced delays to the programme and cost overruns as both agency and contractor sought to reconcile differences. With reliability issues constantly coming up, qualification flight tests were delayed and it became apparent that a completely new system of organisation, procedural activities and the application of test equipment needed a completely new approach.

Key to everything was safety, with the effort targeted toward man-rating the machine (making it safe to carry humans) and machine-rating the man (making sure the human could safely control the machine). With Mercury there was the ever-present concern of operating in a goldfish-bowl environment, with all the associated pressures from remorseless observation and comment from public and media factions alike, usually conducted by people unfamiliar with the programme, its engineering and its essential needs.

Less than a year after it formed, NASA's Space Task Group sought a means of defining and predicting failures and increasing reliability, an effort shared with McDonnell. Taking the growth in statistical quality control during the 1950s that permeated the aircraft industry, this numerical predictor was merged with the science of operations analysis. However, from the start of NASA the senior engineers looked

ABOVE Along with selection of McDonnell to build the Mercury spacecraft NASA set down the requirements for seven astronauts, selected in early April 1959. The men were highly experienced pilots and brought levels of expertise from the Air Force, the Navy and the Marine Corps. From left to right: Malcolm Scott Carpenter, Gordon Cooper, John Glenn, Virgil 'Gus' Grissom, Walter Schirra, Alan Shepard and Deke Slayton. The aircraft behind is a Convair F-106B-75-CO (59-0158) sent to the Military Aircraft Storage & Disposal Center as FN019 on 21 October 1983, thence to become a drone (QF-106.AD275) before disposal to a salvage yard, from where it was rescued and recovered to a trailer park at Ecno Lodge Hotel, Mohave. It is now displayed near the west gate at Edwards Air Force Base, California. *(NASA)*

LEFT Selection of the Little Joe rocket, which had been designed by the Pilotless Atmospheric Research Division of NACA's Langley Aeronautical Laboratory, allowed flight tests to proceed alongside development. The specific launch being prepared here is LJ-1B. *(NASA)*

ABOVE The seven Mercury astronauts pose for a publicity shot with a scale model of the Atlas rocket and spacecraft that would be used for orbital flights. *(NASA)*

to the scientific management of efficiency from the automotive industry. This changed within a year when engineering design was organised around component performance, and while the STG and McDonnell struggled to make that work NASA HQ asked the Air Force for help from its systems engineering people.

The Mercury programme and the safety and quality control criteria it pioneered were the forerunners of the current aerospace M^5 model (man, machine, media, management and mission) operating in a unified and integrated template irrespective of the specifics of the manned programme, be it air or space. Risk management was the key but took the experimental support level to a new height. Previously, manned flight research involved a contractor providing a vehicle for basic experimental research, usually for a specific customer and with an end objective in mind. NACA had been a 'contractor' supporting the development of flight and flying; now, NASA was in charge of experiment, test, qualification and operational management of programmes that it had set itself in a pre-programmed agenda. It was now both consultant and user.

The need to recognise this in the engineering and management philosophy of Mercury weighed heavily on NASA boss T. Keith Glennan and he hired McKinsey and Company to report on its performance in hiring contractors, defining requirements and managing risk.

LEFT Astronaut training for Mercury was more rigorous than for any other subsequent manned space programme. One concern was disorientation in weightlessness. To allow astronauts to practise securing control of a tumbling spacecraft, a fiendish device known as MASTIF (Multiple Axis Space Test Inertia Facility) was installed at the Altitude Wind Tunnel at the Lewis Research Center at Lewis Field, Ohio. Known notoriously as the 'gimbal rig', it was controlled by an operator who could induce complex rotational, spinning or tumbling motions, or combinations of all three. Seated in a plastic chair at the centre, the occupant was tightly strapped down apart from his arms, his right hand being used to practise bringing the device to a halt at a fixed attitude. *(NASA)*

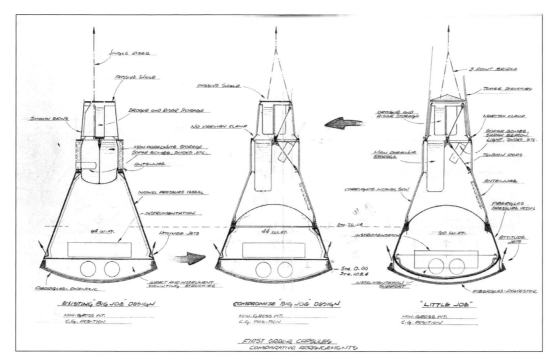

EXISTING "BIG JOE" DESIGN

COMPROMISE "BIG JOE" DESIGN

"LITTLE JOE"

FIRST ORDER CAPSULES
COMPARATIVE ARRANGEMENTS

LEFT As NASA moved toward experimental tests various configurations of capsule were considered, the 'Big Joe' flights being an initial heat shield test on an Atlas rocket. *(NASA)*

BELOW The final configuration shown here was defined in the contract awarded to McDonnell, with an adapter designed to fit the spacecraft to a Little Joe rocket. *(NASA)*

It found that NASA had been negligent in providing adequate work statements, funding provisions and supervision of contractors. The inadequate supervision of contractors was surprising, since NACA had been so strong in its dealings with the Department of Defense. But contractors were now exposed to dual channels of supervision: one from headquarters and another from the relevant field centre.

The agency drew heavily upon industry, to take advantage of a growing inventory of skills and capabilities in a wide range of highly specialised sectors, both in engineering and manufacturing. Clearly, a lot rested on the shoulders of the prime contractor: NASA's reputation, and confidence in this nascent organisation that many doubted was up to the job. For its part McDonnell set up a classic vertical line organisation under manager E.M. Flesh with speciality groups such as stress, aerodynamics, test and others – all horizontal organisations – fed in to the line at appropriate places. About 900 engineering people were involved of whom approximately 500 were directly involved in manufacturing the spacecraft.

RIGHT Flights on an Atlas called for a different form of adapter to mate the spacecraft to the rocket. *(NASA)*

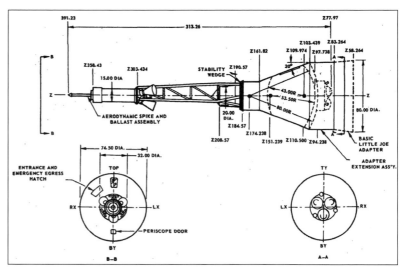

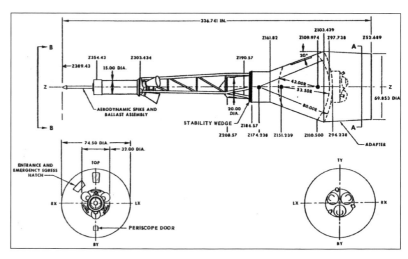

Chapter Three

The Mercury spacecraft (1959–63)

Given its name on 26 November 1958, as America's first piloted spacecraft Mercury was a challenge without precedent. But what emerged pioneered every other spacecraft that followed – not necessarily in shape but in the complexity and evolution of systems, subsystems and countless components. Nothing that followed would have been possible without Mercury, the long-pole on a fast learning curve, a true pioneer in a new age of spacecraft engineering.

OPPOSITE Spacecraft 13, flown on the MA-6 mission, put John Glenn in space – the first American to orbit the Earth. On display at the National Air and Space Museum, Washington DC, it was designed for the astronaut to achieve full authority over its attitude through a complex series of variable modes, from fully automatic to fully manual. *(Via David Baker)*

Structure

The structural requirement of the Mercury spacecraft was defined by its mission: to support a single astronaut for flights in low Earth orbit for up to 28 revolutions and return him safely to splashdown and recovery. Early in the programme it was accepted that each flight required a spacecraft uniquely configured for the specified mission, and, since it was also a development vehicle for spacecraft systems, for manned missions each would be used only once. Nevertheless, the essential dimensions, fabrication and assembly of the primary structure would be common to all spacecraft built.

While the spacecraft could not change its orbit, except for re-entry, it was required to orientate its attitude and to accurately fix its alignment in pitch, roll and yaw for several reasons, the most important being the pointing angle for retrofire, which would determine the angle at which the fight path intersected the atmosphere. Immediately after separating from the launch vehicle it was to turn around so that the astronaut had his back to the direction of travel and pointed 34° apex down, the attitude for retrofire. After firing the retro-rockets at the end of its orbital stay it would realign to a pitch of 14° for entry-interface, the point at which it would re-enter the atmosphere as measured by a 0.05g acceleration sensor.

While the original requirement for Project Mercury had been to produce a spacecraft capable of coming down on water, there had been considerable attention to the possibility of coming down on land. This came about when detailed analyses of escape trajectories displayed a high possibility of coming down on land, which was therefore added to the general requirement of the programme. For a water landing the original specification was for a vertical descent rate of 30ft/sec (9.1m/sec) with a maximum surface wind of 18kt (33.3kph) and waves equivalent to a sea state 4. The original crushable honeycomb structure beneath the astronaut was not adequate to mitigate the deceleration forces or onset rate and it was then that the impact landing bag concept was born.

While counter-intuitive, the structural configuration referred to here is the one used by engineering design teams and assumes the attitude reference to be blunt-end forward (the forebody) and the conical section aft (afterbody). In this orientation, the front is the heat shield and the back is the section supporting the launch escape system. In this reference the astronaut is therefore flying with his back to the direction of travel.

As defined by NASA, the spacecraft had a dish-shaped forebody and an afterbody consisting of a conical mid-section attached to a small cylindrical section known as the recovery compartment. Of conventional semi-monocoque construction, the primary structure was fabricated from titanium. The structure was required to protect the pressurised cabin from high noise levels from the launch vehicle, excessive temperatures from solar radiation on the exterior and from micrometeoroids that might impact the spacecraft. It was also required to provide ease of entry for the astronaut, viewing windows, and an emergency escape route after splashdown.

The launch escape tower (LET) was attached

BELOW This schematic displays the different materials utilised across the structural assembly of the afterbody. *(McDonnell)*

BOTTOM Hat-section stringers supporting the exterior skin and shingles allow for thermal expansion while retaining the integrity of the encapsulated pressure vessel. *(McDonnell)*

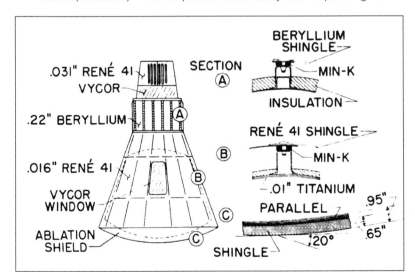

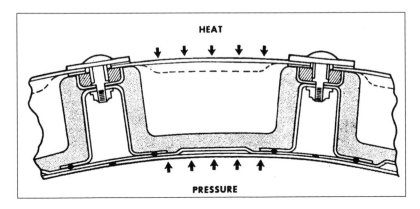

to the cylindrical section of the afterbody and comprised a pylon framework supporting rockets that would carry the spacecraft to safety in the event of an abort. An antenna fairing comprised a cylindrical structure that supported the radio receiving and transmitting antennas. The LET would be jettisoned during the ascent phase when it was no longer required and after an abort; this would expose the antenna section, which would itself be jettisoned to deploy the main parachute subsystem.

From the base of the heat shield to the tip of the launch escape tower the spacecraft had a total length of 24.58ft (7.49m) with a maximum diameter of 6.2ft (1.89m) across the base of the forebody and a minimum diameter of 32in (81.28cm) across the recovery compartment. Without the escape tower the spacecraft had a length of 9.53ft (2.9m) from the base of the heat shield to the top of the stability wedge on top of the antenna fairing. With the retro-package attached the entire spacecraft had a length at launch of 25.9ft (7.89m). In orbit, including retro-package, the spacecraft had a length of 10.96ft (3.34m) and on re-entry a length of 9.59ft (2.92m).

The conditions under which the spacecraft were required to operate were extreme. While exiting the atmosphere on the way to orbit, friction would raise temperatures at the front end to 150°F (65.5°C) while the forward section of the conical spacecraft would experience 700°F (371°C) and the lower circumference of the spacecraft a temperature of 1,300°F (704°C). During peak conditions at re-entry these temperatures would increase, respectively, to 600°F (315°C), 1,000°F (538°C) and 850°F (454°C). The base of the heat shield would experience 3,000°F (1,650°C), but the shock wave standing proud of the blunt forebody would be a blistering 9,500°F (5,250°C), a temperature which would have melted any known material. Between these extremes of ascent and de-orbit, in space the Mercury capsule would experience temperatures of 200°F (93°C) to -25°F (-32°C).

From the outset the design of the capsule was based around two concentric structures: the inner pressure vessel and the outer thermal protective layer. The inner pressure vessel began as trapezoidal sheets almost 6ft (1.8m)

in length with a thickness of 0.01in (0.025cm), commercially available and supplied by Titanium Metals Corporation. These skins were then fusion-welded to form two cones. The inner vessel was fabricated from flat-rolled titanium while the outer vessel was beaded titanium. The two cones were fabricated and tested separately and then nested and spot-welded to hold them in position while the final seam welding was carried out.

Bending, tension and compression loads on the capsule were taken by a framework of commercially pure titanium stringers, hat-sections that were attached to the double-wall pressure vessel by spot welding. A layer of Johns-Manville Thermoflex ceramic-fibre insulation was wrapped around the pressure vessel and up to the outer faces of the hat sections, the tops of which were insulated with a thin strip of Min-K 1301.

Throughout much of the assembly the structural material employed in the fabrication

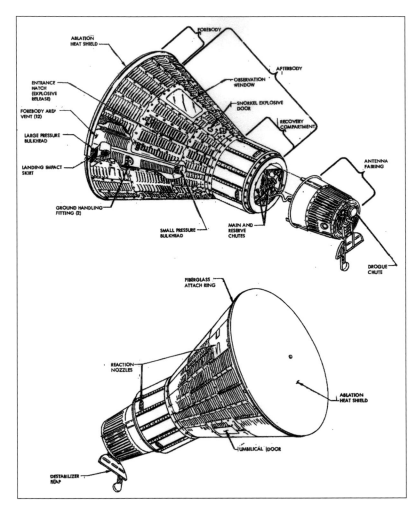

ABOVE **This general arrangement drawing shows the relative relationship of the forebody and heat shield, the afterbody and the recovery compartment with deployable destabiliser flap. The antenna fairing provided a structural support for primary elements of the communications system's small antenna array.**
(McDonnell)

0.01in (0.025cm) titanium welded together for a unitised structure.

Through corrugations or beading of metal in the outer layer of skin, the section of 0.02in (0.05cm) metal became equal in rigidity to a single-sheet thickness of 0.05in (0.127cm), a 150% increase in the efficiency of the structure weight-for-weight. The spacecraft incorporated 45,000in (114.3m) of welding, consisting of 24,500in (62.2m) of seam and butt welding and 20,000in (50.8m) of spot welding. In the production of 20 capsules, total welding exceeded 14 miles (22.5km)!

All seam and spot welding was done open-air, which was unusual since most welding was done in a gas chamber that surrounded the entire welding procedure with an inert atmosphere. McDonnell engineers decided that this was out of the question and designed new fixtures so that all fusion welds were inert-gas shielded, with tungsten arc on both sides of the weld using trailing and backup shields. Chilling of the weld area was also provided to control the temperature of the material. When seen up close, the appearance of this technique was a beauty all its own; this writer personally observed how the weld appeared to be a straight insert of metal set flush between the two titanium skim sheets.

According to tests and qualification trials conducted by McDonnell, these welds were stronger than the original parent metal. Yet spot and seam welding was conducted on standard production equipment employing techniques similar to those used on stainless steels. The ease of use and the range of applications made it a material of choice for the missile industry and the nascent space programmes, Pioneer IV being the first application on a spacecraft or satellite.

A largely unsung benefit from the missile programmes of the late 1950s, materials research had given an unprecedented boost to the industrial production, manufacturing and applications of exotic steels and metal products. Titanium was the big winner. From 6,000lb (2,700kg) of titanium produced in 1948, in 1960 almost 2lb million (907,200kg) was produced for the missile industry alone. This trend would continue and where once the aircraft industry had fed initial developments in the missile and satellite industry, the requirements of the new and exotic manned space projects were turning

of the spacecraft was titanium, then one of the newest metals around and only in commercial production since 1948. Mercury was basically a titanium structure, chosen for its endurance, stiffness and reliability. The use of titanium had the added benefit of a 44% reduction in weight compared to equivalent use of steel. Titanium was also chosen for its ability to retain strength and rigidity at high temperatures. The titanium stringers would experience temperatures of 600°F (315°C) while those on the inner wall would reach 200°F (93°C).

Structural evaluation of the material's rigidity certified the integrity of titanium up to 20g and stiffness characteristics were good despite it having a lower modulus (a measure of elasticity) than steel. Moreover, less density meant greater cross-section and increased stiffness, holding good throughout a wide range of temperatures. In assembling the spacecraft, 'rigidising' or beading also added stiffness to the structure. The cabin walls consisted of two layers of

the tables. Many materials and manufacturing applications would be utilised by the aircraft industry for new generations of military aircraft and that trend would continue.

The first structural shell of a Mercury spacecraft was delivered to Langley on 25 January 1960 and the first capsule for a manned mission arrived at Cape Canaveral on 9 December.

Forebody section

The large dish-shaped, convex forebody – so named because it faced forward most of the time during orbital flight – comprised the heat shield, either ablative or heat-sink, and rings, clamps and associated fittings. It supported the large pressure bulkhead that separated the forebody from the afterbody. The ablative heat shield was constructed of fibreglass shingles laminated to a smooth contour with a final diameter of 74.44in (189cm) and a spherical radius of 80in (203.2cm), engineered to take loads from accelerations during ascent, retro-rocket firing, parachute deployment and aerodynamic loads.

The forebody structure was designed to absorb the impact of striking the water at splashdown and to protect the capsule from the thermal extremes during re-entry. It was secured to the titanium attachment ring riveted to the inner skin of the conical afterbody. This attachment ring carried 48 elongated holes for securing the heat shield and to allow for thermal expansion. The shield was secured to the attachment ring by 24 locking studs alternated with 22 guide studs, with two holes unused beneath the heat shield release mechanism. The retrograde package was secured to the heat shield by three straps and was jettisoned after firing but prior to re-entry.

A series of vents located around the periphery of the forebody between the bulkhead and the heat shield vented trapped gases to the atmosphere, and two toroidal-shaped hydrogen peroxide tanks and six reaction control nozzles, each covered with Min-K thermal insulation, were located in the forebody. The heat shield itself had two functions to perform which were unique to Mercury: protection from the heat of re-entry and its use as a component of an air bag system to cushion landing impact.

The heat shield was designed to detach while the capsule was descending on parachute, and to deploy approximately 4ft (1.22m) below the bulkhead. Supported by a rubber cloth landing impact skirt it would serve as a shock absorber to attenuate the energy felt on striking the water, or the ground if necessary. This area was also used to contain the pneumatic heat-shield release mechanism. After landing, the impact skirt would also serve as a sea-anchor and stabilise the spacecraft on the water, allowing the astronaut to leave either through the side hatch or out the top.

The heat shield was released during descent 12 seconds after ejection of the antenna fairing, which was located above the recovery section. Deployment was initiated by an actuating mechanism that would initiate withdrawal of 24 U-shaped slides to release 24 locking studs together with the heat shield. The impact skirt was attached to the heat shield by a retaining ring containing holes to mate with 219 equally

ABOVE Behind what passed for a cleanroom environment in the late 1950s, a Mercury spacecraft comes together at the McDonnell facility in St Louis, Missouri. This orientation clearly shows the parachutes packaged in the forward section of the recovery compartment. *(McDonnell)*

spaced helical coil inserts in the inner edge of the circular shield at a radius of 34.98in (88.85cm) from the spacecraft's longitudinal axis.

On impact the skirt would fill with air, which would be released through a series of holes in a cushioning effect. A reinforced laminated fibreglass shield over the external surface of the bulkhead would protect it should the heat shield strike the underside of the spacecraft as the impact skirt collapsed. The impact shield was attached to the torus tank support brackets. Sandwiched between the fibreglass shield and the large pressure bulkhead were sections of honeycomb. To prevent the impact skirt from tearing should the capsule be dragged laterally through the water by currents or by the parachute, 24 stainless steel straps were located at equal spacing around the circumference of the skirt, each fitted with a stainless steel cable which would hold fast the deployed heat shield should the skirt be torn away.

The spacecraft's afterbody shingle sidewalls extended beyond the large pressure bulkhead to the heat shield and enclosed the area between the two subsections. The shingles were fabricated from high-temperature alloys by Haynes Stellite Company (no connection to this publisher!). A fibreglass attach ring was bolted to the heat-shield attachment ring that provided a mounting for the spacecraft/launch vehicle adapter. The fibreglass attach ring and the

attachment flange on the adapter were clamped together by a segmented ring. Receptacles for the retro-package, the adapter and the clamp ring electrical and pneumatic connectors were situated under the forebody shingles adjacent to the fibreglass attach ring.

Heat shield

Two types of heat shield were tested for Mercury spacecraft. A beryllium heat-sink shield was employed on Redstone suborbital missions, and an ablative type on orbital flights launched by Atlas. The ablative heat shield was supplied by the General Electric Company and by BF Goodrich. The shield was geometrically a 74.5in (189.2cm) diameter spherical segment with a radius of curvature of 80in (203.2cm). Initially it consisted of two laminates: an ablation laminate 1.075in (2.73cm) thick and an inner structural laminate 0.55in (1.4cm) thick. The ablation laminate was fabricated from concentric layers of fibreglass cloth orientated with the individual layers parallel to the outer surface. Both the ablation and structural laminates were made from a Volan-A finish fibreglass cloth with a 91LD resin.

Following flight tests and arc-jet thermal simulation of the re-entry environment the materials underwent a review that resulted in some changes. In definitive form the cloth was impregnated with CT 37-9X resin, which was a modified phenolic resin developed specifically for the Mercury programme as a result of flight tests and ground simulation of the thermal environment. Resin content of the ablation and structural laminates was 40% and 30% by weight respectively. A circular ring 3in (7.6cm) high made of fibreglass and resin was attached to the back of the heat shield and served to bolt the heat shield to the pressurised compartment of the capsule.

As development tests progressed some changes were made to the formulation of compounds for the two laminates and in the definitive shield the total thickness was 0.95in (2.4cm). The outer laminate, which was known by the engineers as the shingle or ablation laminate, was 0.64in (1.62cm) deep, incorporating separate plies of cloth swept back with respect to the flow direction during the

BELOW Development of the Mercury spacecraft proceeded with the strong support and cooperation of the seven astronauts, a more harmonious and productive exercise than could be inferred from such films as *The Right Stuff*. Here, in Hangar S at Cape Canaveral, astronauts Wally Schirra (left) and Gordon Cooper inspect spacecraft 16, scheduled for MA-8. *(NASA)*

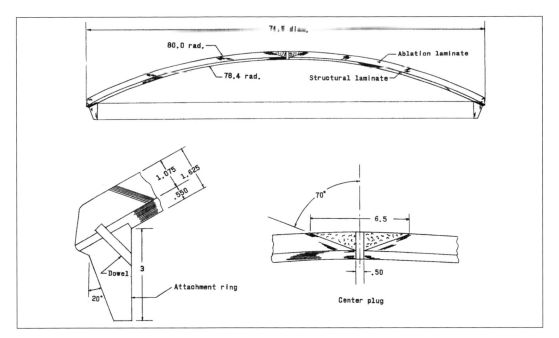

period of heat build-up on re-entry. The angle between the cloth plies and the local tangent to the heat shield was approximately 20°. This type of construction of the outer laminate allowed the gaseous ablation products to escape into the boundary layer without danger of shield delamination. The inner laminate was 0.3in (0.76cm) thick and utilised a parallel layup. This was the structural element of the heat shield and was used to absorb the re-entry air loads and landing impact stresses.

The physical and chemical integrity of the ablative shield was put to the test with the Big Joe flight involving Atlas 10-D and a spacecraft boilerplate launched in 1959. This was the first major test of a piece of technology specifically designed for the Mercury spacecraft.

The ablation shield for Big Joe was instrumented to obtain temperature and char-penetration time histories during flight. Special sensors had been developed by the Missile and Space Vehicle Department of General Electric at their Philadelphia, Pennsylvania, facility. Thirteen of these sensors were distributed around various locations on the shield. Each consisted of one of two types designed to record specific items of measurement. In one, six thermocouples were spaced in depth and in another some 20 pairs of wires were spaced in depth to form a single data intelligence map. The individual pairs of wires utilised the property of the resin, becoming an electrical conductor when charred to complete a circuit between adjacent wires.

Part of the thermocouples in several sensors could not be used due to the limited number of commutator spaces available for telemetering. Three of the available six thermocouples were used in nine of the sensors and all six thermocouples consisted of coaxial chromel-alumel wires insulated from each other and welded together at the hot junction. The 51 thermocouples used in the heat shield were commutated on a single telemetry channel by means of a 60-point switch. The switching rate

BELOW The heat shield doubled as a shock attenuator when deployed with a landing impact bag to cushion the deceleration on hitting the water. Operating like an airbag, on impact it released trapped air through vents in the skirt. (NASA)

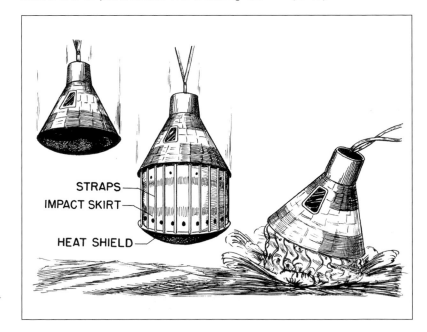

STRAPS
IMPACT SKIRT
HEAT SHIELD

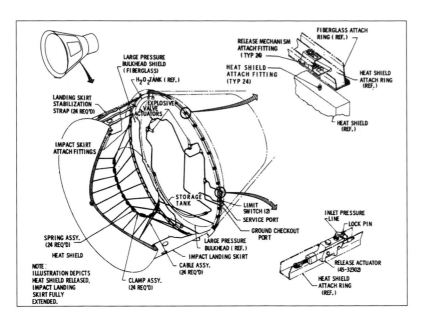

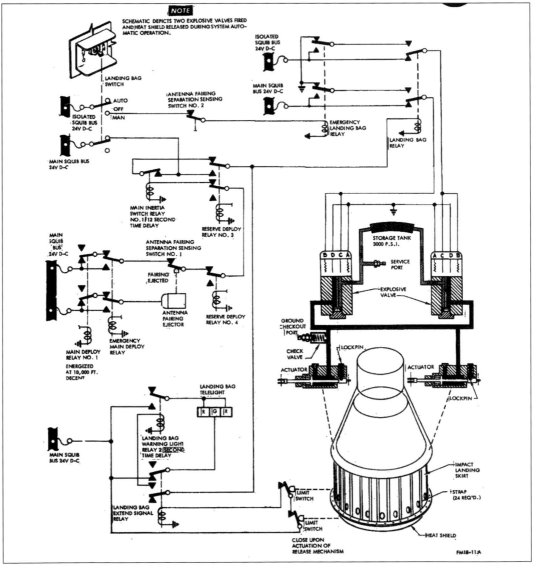

was such that each commutator was sampled about once every 0.62 seconds. Included with the data during each cycle of the switching motor was a major identification pulse as well as the calibrate signals consisting of low, medium and high values of the pre-selected thermocouple temperature range. The maximum temperature of the pre-selected range for the thermocouples was about 2,600°F (1,427°C).

The Big Joe test was only a partial success (see Chapter 4). Nevertheless, the peak heating rate obtained was approximately 77% of that desired and the total heat rate was determined to be 42% of the desired value of 7,100Btu/ft² (22,397W/m²). The only significant damage to the heat shield was a 3in (7.6cm) delamination that occurred near a stagnation point plus

ABOVE Along with insulation and some electrical wiring and delivery lines, hydrogen peroxide propellant for the attitude control thrusters was contained within the space between the heat shield and the adjacent bulkhead. *(NASA)*

RIGHT This diagram displays the electrical circuits dedicated to landing-bag deployment and the discrete lock-outs that prevented premature separation from the forebody of the conical spacecraft. It also shows the interconnectivity with the jettisoning of the antenna fairing and the manual override provided for the astronaut. This schematic represents the system configured for spacecraft 18. *(NASA)*

RIGHT While the landing-impact bag performed well on attenuating the shock of splashdown, it also contributed to unstable conditions on the surface of the sea. This shredded landing bag is acting as a drag-anchor on the spacecraft, keeping the recovery compartment on the waterline. *(NASA)*

a small crack at the edge of a plug and separation around a smaller plug. It was found that delaminations and cracks did not occur below the visible charred portion and did not affect the structural integrity of the shield.

After weighing it was determined that the shield had lost 6lb (2.72kg) and the visible charred region penetrated to a depth of about 0.2in (0.5cm). A discoloured layer extended to a total depth of about 0.35in (0.9cm). It was also determined that charring began at a temperature of approximately 1,000°F (537°C). Specific gravity measurement of the shield indicated that in the charring process only 25% of the available resin content was lost and that surface re-radiation and ablation dissipated about 75% of the total heat load.

Afterbody section

This consisted of a mid-section comprising a pressurised cabin with a small pressure bulkhead adjacent to the recovery section, and a large pressure bulkhead, the latter positioned adjacent to the afterbody heat shield.

The exterior wall of the conical section consisted of an unbeaded inner skin seam-welded to a beaded outer skin with 24 equally spaced longitudinal hat stringers supporting the afterbody heat protection, or radiation, shield. This radiation shield consisted of numerous individual corrugated shingles, secured by bolts to oversize holes for thermal expansion while constraining flutter limits. The shingles were fabricated from René 41, also used on the cylindrical recovery compartment. Thermoflex insulation placed between the outer skin and the radiation shield, bonded over the hat stringers, minimised transfer of thermal loads during re-entry and also reduced noise levels. This also served to protect the pressure cabin from micrometeoroid impacts.

LEFT This assembly shop view shows the aft bulkhead and the associated semi-circular attitude control propellant tank wrapped around one half of the circumference. *(McDonnell)*

BELOW A close-up of the aft bulkhead showing the aluminium honeycomb insulation packed around the electrical wire bundles and pieces of equipment. *(NASA)*

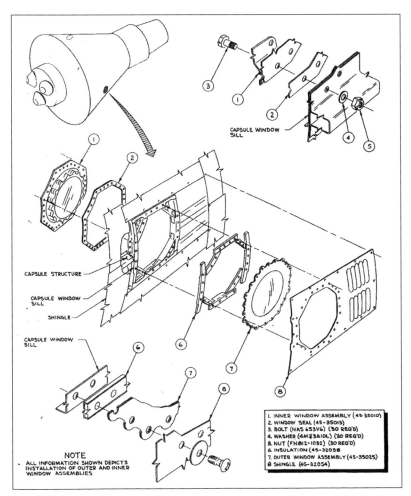

Several issues had been raised during development of the afterbody structure and in particular the heat-protecting shingles. The demands on the structure were high, with considerable noise levels and potential flutter behaviour on the curved corrugated panels at high temperatures. A development test programme before the design of the hardware would have been beneficial, as many programme engineers and managers began to realise when results came in following the test of a boilerplate Mercury on an Atlas flight in September 1959, the first major test in a Mercury Atlas configuration of the heat shield on the base of the forebody and on the shingles covering the afterbody. There was concern in particular about the shingles around the recovery compartment (which see later) and on the antenna fairing attached to that structure.

To accommodate the higher heat levels seen on that flight the shingles on the antenna fairing were replaced with shingles of increased thickness. It was then that the shingles made of the superalloy René 41 were adopted. Originally the heat-protection system for the recovery compartment utilised 0.22in (0.56cm) thick cobalt-alloy shingles on the conical section of the spacecraft. Sufficient doubt had been cast on the original choice of cobalt-alloy for a decision to be made by the STG to replace them with shingles of similar corrugated design, but with the material changed to René 41 and thickness increased from 0.010in (0.025cm) to 0.016in (0.040cm).

Located at the forward end of the conical section, the large pressure bulkhead separated the pressurised compartment from the forebody heat shield and consisted of a combined inner and outer titanium skin, beaded and seam-welded together. The bulkhead was reinforced with two vertical channels located centrally and spaced for a structural attachment point for the astronaut's couch. The inner skin of the bulkhead supported honeycomb shelves for installation of equipment. The outer flange of the bulkhead was bolted to the conical section's inner attachment ring as well as to the inner skin. Vents were machined in to the large pressure bulkhead to allow for the release of battery vapours and exhaust steam from the environmental control system.

The small titanium pressure bulkhead separated the cabin from the recovery section and was a structural support for the aft conical section. It was seam-welded to the inner skin of the conical section and bolted to the hat section flanges. It also supported a small hatch that was actuated internally as the primary means by which the astronaut would emerge after landing. Dish-shaped, the hatch was constructed of a beaded aluminium skin spot-welded to an inner skin reinforced with Z-shaped members. The outer flanged edge of the hatch fitted into the small pressure bulkhead and was held in place by a retaining ring, which was expanded by raising the actuator handle, wedging the ring between the sill and the flanged edge of the hatch to create a seal.

Three other penetrations of the conical pressure vessel each contained a periscope, a viewing window and an astronaut access hatch. The main observation window was designed in as a response to pressure from the astronauts. It provided a field of view 30° in the horizontal plane and 33° in the vertical. It was not installed on early unmanned capsules or on spacecraft 7, flown by Alan Shepard on MR-3, MR-4 being the first manned flight with this window. The retractable periscope housing door was located on the bottom (Earth-facing side) of the conical afterbody which encapsulated the periscope's lower lens flange and also the ground umbilical receptacle. Mechanically linked to the periscope housing, the door automatically opened and closed with extension or retraction of the periscope as commanded by the astronaut.

The single main window in the conical afterbody was centrally located above the main instrument panel and directly above the astronaut's viewing area. It had a length of 19.5in (49.5cm), a maximum width of 12in (30.5cm) at its widest end (closest to the forebody) and a width of 9.5in (23.75cm) at the opposite end.

The observation window consisted of an inner assembly with three glass panes and a fourth Polaroid image suppression filter pane to eliminate reflections and secondary images. Each inner pane was individually sealed to provide a pressure seal between each of them, but not the outer pane. The light transparent Polaroid filter layer was the innermost of the four inner panes. The window area was provided with two half-doors and separate filters to shut out the light if necessary. The outermost pane carried lateral reference sight lines on its inner and outer surfaces as required by the mounting angle and the fixed optical reference point. The two inner panes were each of 0.34in (0.86cm) tempered glass and the outermost pane on the inner assembly was of 0.17in (0.43cm) Vycor glass.

The outer window conformed to the curvature of the capsule and consisted of a single pane of 0.35in (0.89cm) Vycor glass sealed in a titanium frame attached to the outer skin of the capsule. The outer pane was required to have a grade 3N optical fidelity in relaxed areas and 2N in two critical areas possessing an optical deviation of 2" of arc. The critical areas were reference sight lines on the inner window assembly as noted above. The set of lines near the base of the trapezoidal window provided a line-of-sight reference for viewing the Earth's horizon compatible with

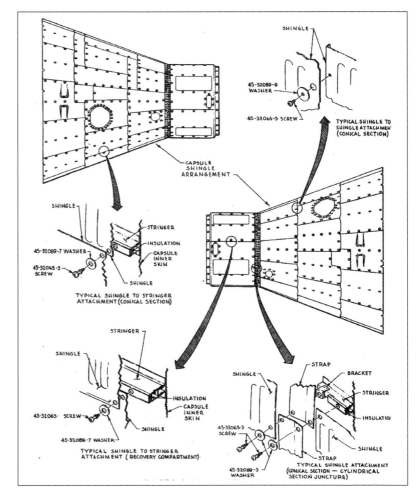

the retrograde orbit alignment of -34° and the second set provided a reference alignment with the capsule's attitude orientation of -14.5° from the horizontal, heat shield up. The outer window was sealed separately from the inner window.

The panes for the inner window assembly were mounted in individual support rings, each sealed with gaskets on the upper and lower surfaces and held rigid by spacers around the edge inside the ring. The inner surface of the outer window pane and both surfaces of the inner window assembly panes were coated with a single layer of magnesium fluoride (MgF_2) film to inhibit thermal radiation entering the pressurised cabin area.

Special sets of covers and filters were provided to protect the astronaut from solar radiation and the lighting effects of the boundary layer during re-entry, when a plasma sheath would build up around the spacecraft. The cover comprised a double door of aluminium alloy with honeycomb core and was shaped to the angular frame of the trapezoidal window. Each half-door was hinged on opposite sides of the window frame with a latching mechanism and handle, retained in the open position by side latches. The filters consisted of a right and left Plexiglas panel, 0.08in (0.2cm) thick, hinged on the outboard edges, with a rubber sealing strip around the mobile edges and held in the closed position by a spring-loaded latch above and forward of the astronaut's head. When pulled, the latch released the filters to the open position where they could be latched on each side with the cover assembly doors. The filters were red to ease the astronaut's adaptation to night and day.

An extended mirror was provided with the cover assembly, located on the lower end of the inner window assembly and mated with the filter sealing strips. The mirror itself was of aluminium alloy with a reflecting surface. It was adjusted by means of a ring handle when an extended view of the horizon was required. The astronaut also had a special pole, secured by line to the inside of the capsule, for operating the window cover and filter assembly.

The astronaut ingress/escape hatch was located on the right side of the spacecraft as viewed by the astronaut in his couch, and because of the conical form of the afterbody it was trapezoidal in shape and was constructed

in a similar fashion to the basic afterbody proper. The hatch was 25.5in (64.77cm) high, 28in (71.1cm) wide at the sill, and 20in (50.8cm) wide at the top. It was not permanently attached to the spacecraft and was installed after the astronaut had been seated in the pressure cabin. Because the hatch was also designed to provide a means of emergency escape after landing, it incorporated an explosive release system through a detonator activated from inside the capsule on all but the first manned flight.

The hatch consisted of an inner and an outer skin, seam-welded and reinforced with hat stringers. On the inside attachments were provided for a waste container and an astronaut's knife. In an emergency an explosive charge located in the hatch sill would be triggered by an initiator located in the upper aft corner and linked to an internal release control. The hatch was bolted in position prior to launch with two corrugated shingles placed over the hatch but not attached to the shingles on the capsule. The bolts were inserted through the hatch sill and threaded into the sill of the capsule. The hatch seal itself consisted of a magnesium gasket with inlaid rubber to form a tight seal when bolted in position.

To trigger the hatch the astronaut first removed the cap from the initiator and the safety pin from the plunger. On depressing the plunger two spring-loaded firing pins would strike the explosive charge percussion caps and detonate the charge. This released the hatch by breaking the 70 attachment bolts through four parallel strands of explosive charge. Two strands of explosive powder were located beneath the bolt heads on each side of the bolt shank. All four strands were ignited from each end simultaneously to ensure redundancy. The explosive charge propelled the hatch some distance from the capsule, thereby allowing the astronaut to escape.

An exterior release handle allowed recovery personnel to release the hatch in the event the astronaut was disabled. Activation retention springs secured by pip pins were attached to the inner face of the hatch so that it would not strike recovery personnel should the plunger be inadvertently depressed from the inside. Two pressure valves in the hatch allowed ground personnel to pressurise the cabin and to carry out purging operations during ground checkout procedures.

Spacecraft 2, 3 and 4 contained two circular side windows but not an observation window since these three spacecraft would not be manned. One window was located in the upper left side as viewed from the crew station, the other in the right side for Earth photography with a 70mm camera installed. These windows were also incorporated into MR-3, flown by Alan Shepard, which did not have an observation window, but not on any other manned flight.

One other orifice was installed: an exploding

ABOVE LEFT
Afterbody alignment in the sheet metal shop as a boilerplate is prepared for completion. *(NASA)*

ABOVE **A production line of Mercury spacecraft with the forebody and heat shield section displaying the box structure on the inner face to which the couch will be attached in the pressure cabin.** *(McDonnell)*

ABOVE This boilerplate spacecraft reveals just how small the Mercury really was, scaled by the personnel operating test equipment. *(McDonnell)*

BELOW Assorted cabin equipment was designed for specific missions, including crewman simulators for unmanned flights, primate enclosure for chimpanzee missions and an astronaut couch for manned operations. *(McDonnell)*

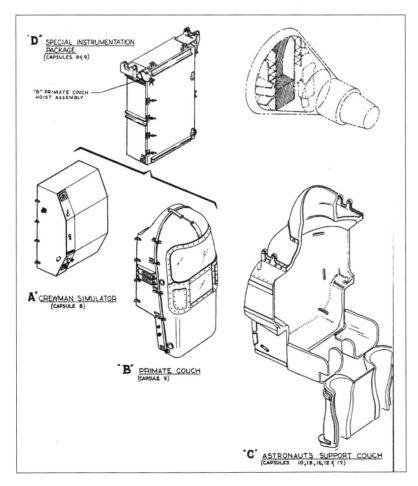

door released after landing to expose the snorkel which would ingest air to the cabin, drawn in by the environmental control system and activated by a valve after the cabin inlet valve opened.

Two auxiliary hoist fittings were also attached to opposing sides of the conical afterbody for ground handling attachment points, removed before flight. An explosives door for the snorkel was provided in the shingles between the small pressure bulkhead and the juncture between the conical afterbody and the cylindrical recovery section.

Combined, the forebody and afterbody had a total length of 5.63ft (1.71m) to the base of the recovery compartment.

Recovery compartment

A functional element of the afterbody, the recovery compartment consisted of a cylindrical section 32in (81.28cm) in diameter and 22.75in (57.78cm) in height. It contained the landing parachutes, the recovery aids and Reaction Control System (RCS) nozzles for attitude control. The structure was formed from a titanium skin reinforced with longitudinal hat stringers and covered with external shingles to protect it from excessive heat on re-entry. Layers of Thermoflex insulation were placed between the stringers and the shingles that comprised individual panels bolted to the stringers with sufficient tolerance for thermal expansion and contraction. The RCS thrusters were situated at 90° intervals between the inner skin and the external shingle installation.

Internally, the recovery compartment was separated into right and left halves. The left section contained the various recovery aids, electrical wiring and items of plumbing and RCS propellant lines. The right side housed a fibreglass container that was structurally separated into two sections for the main and reserve parachutes. After landing, this container could be released by the astronaut to allow him to egress the spacecraft. A separate stability wedge was located on top of the recovery compartment as a mounting plate for the antenna section located above.

The antenna fairing was cylindrical in shape with a modest taper, 24in (60.96cm) high, and

20in (50.8cm) in diameter. In addition to housing the primary receiving and transmitting antenna it also contained the pitch and roll horizon scanners and had a titanium structure covered with René 41 shingles. The skin had an 8in (20.3cm) deep 'window' around the outer base where it attached to the stability wedge on top of the recovery compartment. This consisted of a silicone base with fibreglass insulation, Vycor glass and three Teflon strips. Aligned with these strips were three laminated fibreglass guides attached to the antenna fairing shingles. Together, these prevented damage to the antenna fairing when the launch escape tower was jettisoned. An aluminium bi-conical horn was mounted at the base of the fairing and an electrical insulator with lock-foam was mounted above the horn to further improve insulation. A pitch horizon scanner was attached to the top of the antenna fairing and a roll horizon scanner was mounted at the side in line with the pitch scanner.

The fairing itself was structurally mounted to a mortar gun assembly located in the recovery compartment, with a steel post used as a guide when the fairing was ejected during descent. The attachment was secured by three index pins and six support clips in the lower mating flange and aligned with three bolts and six brackets in the recovery compartment mating flange. Three cables attached the drogue parachute risers to the fairing when the parachute system was deployed.

For re-entry and for abort situations, a destabiliser flap was attached to the top of the antenna fairing on the side opposite the pitch horizon scanner. The flap would be used as an aerodynamic rudder to properly align the spacecraft to re-entry attitude with respect to the atmosphere. It was spring-loaded and held against the antenna fairing up to the point of launch escape system jettison by a nylon cord routed through two reefing cutters installed within the antenna fairing. When the escape tower was jettisoned the cutters severed the cord and released the loaded flap to the deployed position, and when the spacecraft

BELOW The observation window had various glass panes and a filter shade and was specifically designed as an aid to navigation and attitude orientation. The configuration of various optical assets changed over the course of the programme, and on the last manned Mercury flight the window would be used to support a camera mounting. *(McDonnell)*

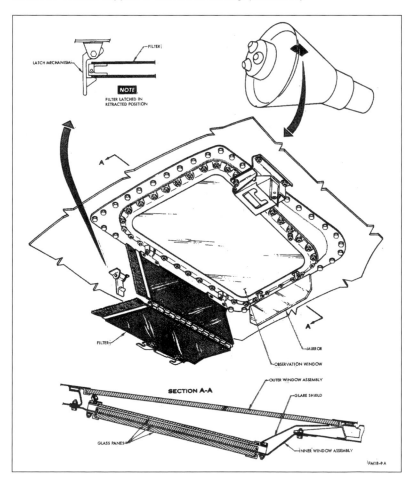

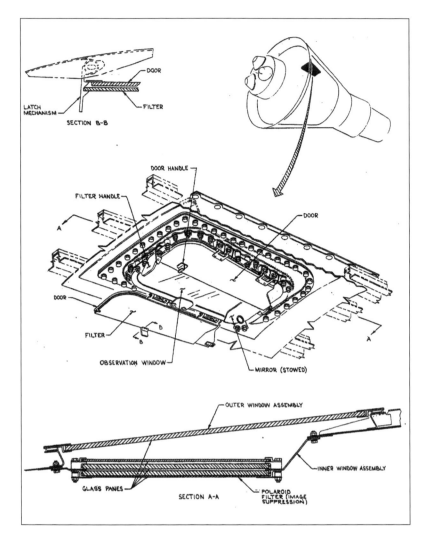

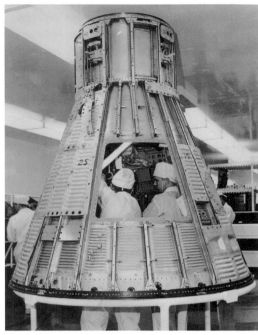

ABOVE Fitting out the spacecraft required much of the equipment and systems to be installed before attachment to the pressure bulkhead and the heat shield, after which the limited dimensions of the hatch limited access to the extremely confined interior. *(NASA)*

LEFT Structural configuration of the observation window with door latches and filter handles. *(McDonnell)*

RIGHT The hatch itself was only just large enough to allow a suited astronaut of limited height to enter, the exercise taking some time due to the large number of belts to be secured and equipment to avoid. Here, Gordon Cooper is egressing after his mission. *(NASA)*

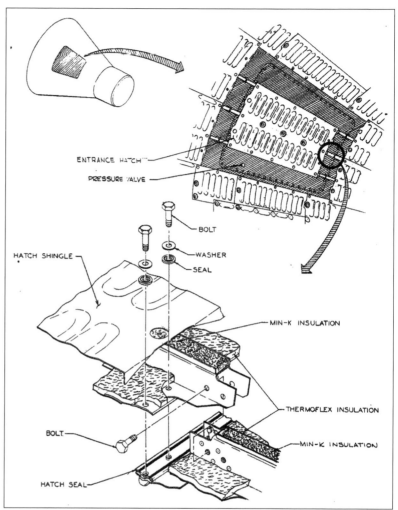

had descended to 10,000ft (3,048m) the fairing was jettisoned from the spacecraft by the firing of the mortar gun.

The barostat which energised the main deploy relay to jettison the antenna fairing started a sequence that closed the antenna fairing sensor switch, sending 24V DC current to the main inertia relay switch. Twelve seconds later this relay sent direct current to the heat shield release limit switches and energised the landing bag relay. This ignited the two heat-shield explosive squib valves. When these fired they opened a 3,000psig (20,685kPag) nitrogen flow to the two shield release actuators that move the shield away from the forebody structure. Activation of the release mechanism also applied power to the landing bag extension signal relay, directing power to the green landing bag telelight in the cockpit, which confirmed completion of the deployment cycle. The actuator was locked in the fully deployed condition by a spring-loaded pin. Disarmed in orbit, this cycle was armed during re-entry by the pilot placing the landing bag switch in the 'AUTO' position.

If the mechanism should fail to actuate, the two limit switches would remain open and the landing bag relay light would energise within two seconds and direct power to the red landing bag telelight, which told the astronaut that it was an unsafe condition. Placing this switch to the 'MAN' position sent power to the

mechanism to release the heat shield, and this would close the limit switches to illuminate the green light.

Even something as simple as the antenna fairing produced design headaches unimagined at the first design stage. Ejected by mortar at an altitude of 10,000ft (3,048m), the simulated firing of test rounds showed problems with the structural members holding it in place. An improper simulation together with the fact that the redesign of the antenna fairing had resulted in increased weight caused some fasteners to fail as well as a main structural web in the cylindrical section during static firing trials. The weakness was remedied by increasing the size of the fasteners and the thickness of the web. While simple from an engineering standpoint, the consequences rippled through the entire structure of the spacecraft and had consequences in added weight and a different ejection velocity.

ABOVE LEFT Seen through the open hatch is the instrument display panel and its associated electrical wire bundles in the pressurised cabin. *(NASA)*

ABOVE Insulation and pressure valve on the standard hatch with bolts and a dedicated shingle overlay to preserve the thermal integrity of the capsule's structure. *(McDonnell)*

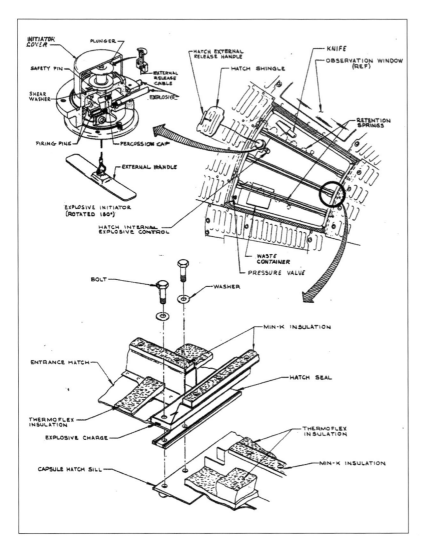

Labels in figure:
INITIATOR COVER
PLUNGER
SAFETY PIN
SHEAR WASHER
FIRING PINS
PERCUSSION CAP
EXTERNAL HANDLE
EXPLOSIVE INITIATOR (ROTATED 180°)
HATCH INTERNAL EXPLOSIVE CONTROL
EXTERNAL RELEASE CABLE
EXPLOSIVE
HATCH EXTERNAL RELEASE HANDLE
HATCH SHINGLE
KNIFE
OBSERVATION WINDOW (REF)
RETENTION SPRINGS
WASTE CONTAINER
PRESSURE VALVE
BOLT
WASHER
MIN-K INSULATION
ENTRANCE HATCH
HATCH SEAL
THERMOFLEX INSULATION
EXPLOSIVE CHARGE
CAPSULE HATCH SILL
THERMOFLEX INSULATION
MIN-K INSULATION

ABOVE The explosive hatch showing structural detail and the plunger required to trigger the jettisoning process, also connected to an external handle for access by attendant personnel should that be required. *(McDonnell)*

Launch escape tower

The Mercury escape system was designed to carry the astronaut and his spacecraft to safety in the event of the launch vehicle running amok. It consisted of a rocket to propel the capsule away from a potential fireball – on the pad or during ascent – and another to carry it away after it was no longer needed. But the design concept was not without its detractors, and with justifiable reason as it turned out.

Initially McDonnell had proposed a cluster of eight separate rockets attached to a fin adapter but the STG preferred Max Faget's idea of a single motor exhausting through three separate canted nozzles. Tests began with a spacecraft mock-up being placed on a pad at Wallops Island where the PARD team conducted a simulated pad abort. On 11 March 1959 the

escape system fired and carried the spacecraft into the air, fully under control for the first 50ft (15m) before it began to pitch through several rotations and came down some way from the shore. Analysis of the wreckage showed that a graphite insert from one of the rocket nozzles had come out, causing asymmetric thrust. On 28 March the alternative was looked at again but the added complexity was felt to increase the risks of asymmetric thrust and the STG adhered to the original concept.

Designed to carry the solid propellant escape and jettison motors, the launch escape tower consisted of a lattice structure for maximum rigidity and light weight. The pylon itself was of triangular cross-section constructed from SAE 4130 steel tubing and had a length of about 10ft (3m). The base was bolted to a steel flanged attachment ring on the top of the recovery compartment and was covered with insulation. The structural tubing of the lattice tower contained electrical cables and lines attached to connectors at the base, for operation of the escape tower and/or jettisoning of the motors. The 4ft (1.2m) long escape motor was bolted to the top of the pylon and the jettison motor was located inside the nest of three exhaust nozzles.

The pylon was attached to the capsule by clamping its attachment ring to the recovery compartment by a chevron-shaped and segmented clamp ring. Explosive bolts placed the clamp ring segments in tension, and were fired to separate the clamp ring from the capsule when the pylon was jettisoned during normal flight. In the event that the abort sequence carried the spacecraft away from the launching rocket the jettison motor would fire to remove the pylon from the spacecraft, exposing the recovery compartment and following a conventional recovery cycle using drogue and main parachutes in the same manner as for a normal descent from orbit.

The three explosive bolts which held the clamp ring had dual ignition and connected the ring segments in a similar manner to that of the spacecraft/launch vehicle adapter, albeit being much smaller in size. The external section of the clamp ring carried a heat shield to protect it from the heat encountered during the normal launch phase ascending through the

A view of the structural geometry of the forward escape hatch in the apex of the pressure dome situated immediately below the recovery compartment, together with the astronaut-activated release handle for egress from the top of the spacecraft. *(NASA)*

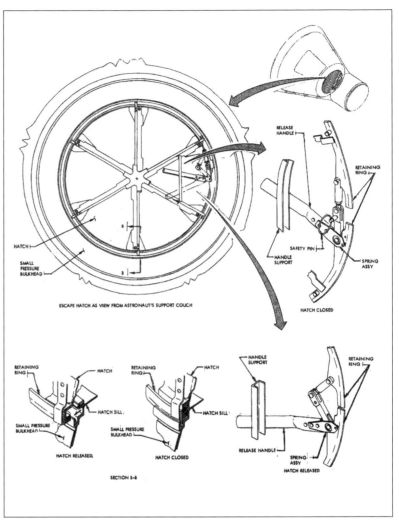

atmosphere. This carried a layer of Thermoflex bonded to the exterior of the shield and the entire assembly was screwed in place. The stability wedge attached to the top of the recovery compartment helped stabilise the spacecraft. Six cable straps were bolted to the pylon structure and the stability wedge to hold fast the clamp ring when the explosive bolts were fired, thus ensuring a clean separation of the tower.

Initial estimates of aerodynamic heating indicated only moderate levels of thermal impact on the various elements, which is why conventional 4130 steel was chosen. However, subsequent and more refined calculations pointed to certain areas having very intense and localised heating and indicated that temperatures of the horizontal members and at the joints of the open truss structure would reach as high as 1,600°F (871°C) during ascent. The remedy was fibreglass insulation added to the horizontal members.

FAR LEFT Spacecraft No 2, which flew on the first Redstone mission, is prepared for test at the NASA Lewis Research Center. Note the sheet metal lower afterbody skirt. *(NASA)*

LEFT A close-up view of spacecraft 11, which was flown by Gus Grissom on MR-4, displays the recovery compartment, topped by the antenna fairing that was jettisoned during descent to expose the parachutes. *(NASA)*

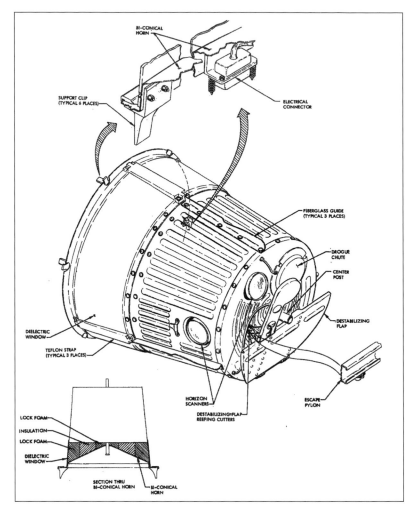

BI-CONICAL HORN

SUPPORT CLIP (TYPICAL 6 PLACES)

ELECTRICAL CONNECTOR

FIBERGLASS GUIDE (TYPICAL 3 PLACES)

DROGUE CHUTE

CENTER POST

DESTABILIZING FLAP

DIELECTRIC WINDOW

TEFLON STRAP (TYPICAL 3 PLACES)

HORIZON SCANNERS

DESTABILIZING FLAP REEFING CUTTERS

ESCAPE PYLON

LOCK FOAM
INSULATION
LOCK FOAM
DIELECTRIC WINDOW

SECTION THRU BI-CONICAL HORN

BI-CONICAL HORN

Launch vehicle adapters

The Mercury spacecraft was fired into space by one of two launch vehicles, each unique in the spacecraft/launch vehicle adapter required. The ballistic flights were flown by the Redstone rocket, which had a diameter of 70in (177.8cm), while the Atlas rocket employed for orbital flights had a top end diameter of 69.8in (177.3cm). Despite the similarity of diameter, each launch vehicle had unique adapter configurations. Essentially, the spacecraft/ adapter interfaces were the same but the adapter/launch vehicle interfaces were unique to the specific vehicles.

The Redstone adapter was a shallow, slightly tapered structure with a cylindrical shape to adapt the 70in diameter of the stage to the 74.5in (189.2cm) diameter of the Mercury spacecraft. It was 15in (38cm) deep and consisted of a semi-monocoque construction fabricated primarily from titanium. It comprised four panels, butt-welded to form the skin with vertical hat sections welded to them internally for structural reinforcement. A steel flanged ring was riveted to the bottom outer surface of the adapter and incorporated holes to allow the attachment of the adapter to the rocket with explosive bolts, which were wired separately for redundant ring cutting.

ABOVE The antenna fairing with destabiliser flap in closed position. *(McDonnell)*

RIGHT A comprehensive display of the separation devices required to get the spacecraft away from the launch vehicle and to release the launch escape tower during an uneventful ascent. *(McDonnell)*

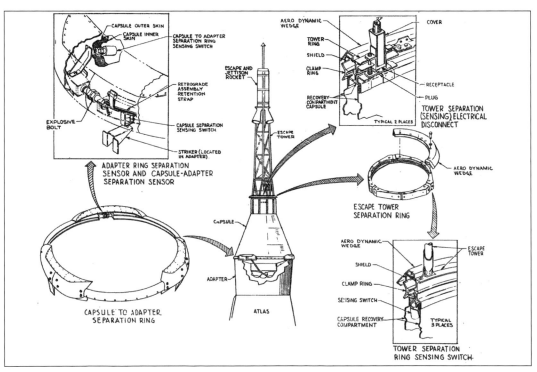

CAPSULE OUTER SKIN
CAPSULE INNER SKIN
CAPSULE TO ADAPTER SEPARATION RING SENSING SWITCH
RETROGRADE ASSEMBLY RETENTION STRAP
EXPLOSIVE BOLT
CAPSULE SEPARATION SENSING SWITCH
STRIKER (LOCATED IN ADAPTER)

ADAPTER RING SEPARATION SENSOR AND CAPSULE-ADAPTER SEPARATION SENSOR

ESCAPE AND JETTISON ROCKET
ESCAPE TOWER
CAPSULE
ADAPTER
ATLAS

CAPSULE TO ADAPTER SEPARATION RING

AERO DYNAMIC WEDGE
TOWER RING
SHIELD
CLAMP RING
RECOVERY COMPARTMENT CAPSULE
COVER
RECEPTACLE
PLUG
TOWER SEPARATION (SENSING) ELECTRICAL DISCONNECT
TYPICAL 2 PLACES

AERO DYNAMIC WEDGE
ESCAPE TOWER SEPARATION RING

AERO DYNAMIC WEDGE
SHIELD
CLAMP RING
SENSING SWITCH
CAPSULE RECOVERY COMPARTMENT
ESCAPE TOWER
TYPICAL 3 PLACES
TOWER SEPARATION RING SENSING SWITCH

Riveted to the top outer surface of the adapter, an aluminium flanged ring mated with the fibreglass attachment ring on the forebody of the spacecraft. The aluminium ring was slotted at 120° intervals to provide clearance for the retrograde assembly attachment straps. Three metal striker brackets were riveted internally to the adapter skin at 120° intervals, depressing separation sensor switches attached to the lower end of the retrograde attachment straps. The spacecraft was attached to the upper ring of the adapter via a chevron-shaped segmented clamp ring over the mated flanges of the spacecraft attachment ring. Electrical connections were provided at two plug interface templates for power between the adapter and the spacecraft.

The Atlas launch vehicle adapter was a tapered cylinder with a lower diameter of 69.8in (177.3cm) and a top diameter of 74.5in (189.2cm), a flared form more noticeable visually if only because of the greater depth of 50.75in (128.9cm). Of semi-monocoque construction, the adapter consisted of an outer corrugated titanium assembly, riveted and seam-welded to an inner titanium skin reinforced with two titanium support rings riveted between the ends of the cylinder. The inner surface of the bottom circumference supported a riveted flange ring that was perforated with bolt holes to allow attachment of the adapter to the top of the Atlas rocket. The inner surface of the top circumference was an aluminium flanged attachment ring that mated to the spacecraft's forebody fibreglass attachment ring.

The capsule/adapter clamp ring secured the capsule until the ring was separated by means of explosive bolts. It consisted of three chevron-shaped segments, as with the Redstone adapter, designed to mate with the flanges on the spacecraft attachment ring in the same manner. The same striker brackets were fitted with the same function. The exterior of the clamp ring was covered with a heat shield in the form of three equal fairing segments covering the adapter clamp ring and located directly above the explosive bolts. The top section was fabricated from aluminium and the bottom one from titanium. Again, the interior was insulated with Thermoflex.

Six cable straps were bolted to the adapter

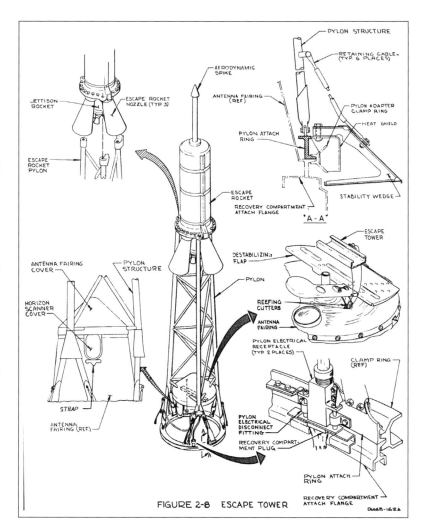

FIGURE 2-8   ESCAPE TOWER

clamp ring, retaining it when the bolts fired. Clamped around the interior of the adapter, an electrical cable was connected to each of the clamp ring explosive bolt chains, fixed to two receptacles in the spacecraft forebody and two similar receptacles on the Atlas. A pneumatic line was also connected to one end of the explosive bolt and to a quick disconnect in the structure of the forebody.

Redstone and Atlas launches were also flown manned, so a man-rating criteria was applied to the adapters fitted to these launchers. Little Joe was somewhat different in that these rockets, designed by PARD themselves, only launched unmanned capsules or boilerplates with shape, size and mass simulation and with a different range of equipment in the spacecraft according to the requirements of the test.

Two versions of the Little Joe adapter were available: one for tests of the launch escape system without retrograde package and one

ABOVE The launch escape tower and associated escape motor and jettison motor was a complex piece of equipment. The launch weight of a typical Mercury spacecraft was reduced by 30% on the way to orbit, most of it in the escape system shown here which weighed in at more than 1,100lb. *(NASA)*

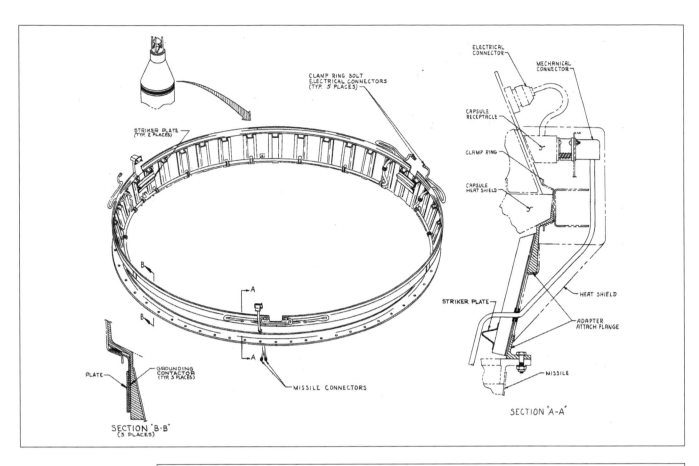

STRIKER PLATE
(TYP. 2 PLACES)

CLAMP RING BOLT
ELECTRICAL CONNECTORS
(TYP. 5 PLACES)

ELECTRICAL
CONNECTOR

MECHANICAL
CONNECTOR

HEAT SHIELD

CAPSULE
RECEPTACLE

CLAMP RING

CAPSULE
HEAT SHIELD

STRIKER PLATE

HEAT SHIELD

ADAPTER
ATTACH FLANGE

MISSILE

SECTION "A-A"

PLATE

GROUNDING
CONTACTOR
(TYP. 3 PLACES)

MISSILE CONNECTORS

SECTION "B-B"
(3 PLACES)

ABOVE There were two separate adapters provided for the Mercury spacecraft, the smaller of which was that to clamp the capsule to the Redstone launch vehicle. *(McDonnell)*

RIGHT The much larger adapter and clamp ring for the Atlas launcher, its extended length providing a suitably sized void inside to contain the Mercury retrograde package. *(NASA)*

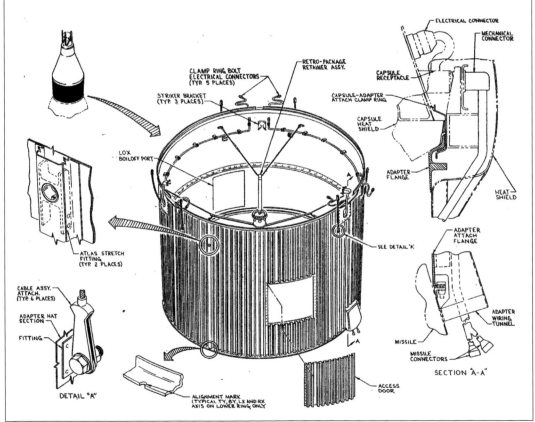

CLAMP RING BOLT
ELECTRICAL CONNECTORS
(TYP. 5 PLACES)

RETRO-PACKAGE
RETAINER ASSY.

STRIKER BRACKET
(TYP. 3 PLACES)

ELECTRICAL CONNECTOR

MECHANICAL
CONNECTOR

CAPSULE
RECEPTACLE

CAPSULE-ADAPTER
ATTACH CLAMP RING

CAPSULE
HEAT
SHIELD

LOX
BOILOFF PORT

ADAPTER
FLANGE

HEAT
SHIELD

ATLAS STRETCH
FITTING
(TYP. 2 PLACES)

SEE DETAIL "A"

ADAPTER
ATTACH
FLANGE

CABLE ASSY.
ATTACH.
(TYP. 6 PLACES)

ADAPTER HAT
SECTION

FITTING

ADAPTER
WIRING
TUNNEL

MISSILE

MISSILE
CONNECTORS

SECTION "A-A"

ACCESS
DOOR

DETAIL "A"

ALIGNMENT MARK
(TYPICAL TY, BY, LX AND RX
AXIS ON LOWER RING ONLY

Diagram labels:

HATCH EXPLOSIVE INITIATOR (ORBITAL CAPSULE)

HATCH EXTERNAL EXPLOSIVE CONTROL (ORBITAL CAPSULE)

ENTRANCE HATCH INTERNAL RELEASE HANDLE (CAPSULES 5 AND 7)

EXTERNAL RELEASE HANDLE SOCKET (CAPSULES 5 AND 7)

ENTRANCE HATCH

LOWER WINDOW (CAPSULES 5 AND 7)

ORBITAL SPACECRAFT

UPPER WINDOW (CAPSULES 5 AND 7)

ORBITAL CAPSULE WINDOW

PERISCOPE

with retro-rocket motor assembly. The basic adapter transitioned the 80in (203.2cm) diameter of the Little Joe rocket to the Mercury spacecraft forebody. Of conventional semi-monocoque construction, it was fabricated from aluminium, steel and magnesium and was 25in 63.5cm) deep. A cylindrical adapter extension assembly, a constant 20in (50.8cm) deep, was an option fitted to the top of the truncated lower section to increase the internal void and provide space for the retrograde package when that was required for the test objectives.

The basic adapter and its extension skirt weighed 914lb (414.5kg), the aluminium extension skirt consisting of a machined structural frame, an adapted Atlas spacecraft fitting and a top adapter ring provided by NASA. As with Atlas and Redstone adapters, McDonnell provided the clamp ring assembly which consisted of the three segmented sections joined by three explosive tension bolts. Two of the bolts could be initiated electrically from either end of a dual electric system, as with the other two vehicle adapters.

Electrical power system

The electrical system for the Mercury spacecraft consisted of three main batteries, two standby and emergency batteries and an isolated battery which supplied currents at 6, 12, 18 and 24V DC through control circuits to the high-priority and low-priority distribution busses. The low distribution circuit could be switched

ABOVE LEFT The Atlas adapter, with launch vehicle electrical and instrumentation leads in conduits to the spacecraft together with the service mast umbilical connections. *(NASA)*

ABOVE Variations and design differences between suborbital and orbital spacecraft highlight the unique nature of individual capsules adapted for specific roles and instrumented with systems unique to mission requirements. *(NASA)*

off to conserve power; external DC power was supplied through a diode panel before launch. When spacecraft power was on, an ammeter indicated bus current levels, and the voltage could be read off the DC voltmeter when either external power or battery power was supplying the busses.

With the exception of the abort control circuit, which incorporated a solid converter, direct current electrical loads were supplied through fuses. Some DC circuits had two fuses which allowed switching between 'normal' and 'emergency' positions and circuits that had a solid conductor in place including the emergency spacecraft separation control, tower separation control, emergency main parachute deploy, retro-package jettison control, retro-package manual control, reserve parachute deploy and emergency reserve parachute deploy.

Alternating current was obtained via two main inverters and one standby inverter, transforming battery voltage to 115V, 400-cycle, single-phase AC distribution busses.

RIGHT The electrical systems layout for a typical spacecraft shows the installed location of the batteries to the left and right of the couch and adjacent to the bulkhead. This layout, specifically for spacecraft 13 for the MA-6 mission, also shows the associated displays and selector controls on the left side of the pressure cabin and on the main display panel. *(NASA)*

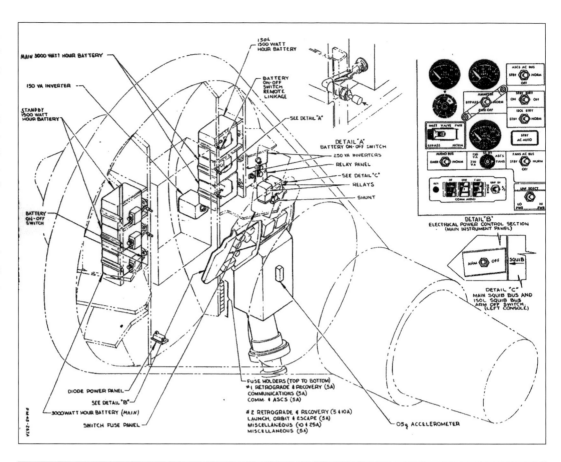

RIGHT The layout for the electrical systems changed by varying degrees dependent on the specific mission assigned to each spacecraft. The configuration shown was for spacecraft 16 (MA-8) and 18 (MA-7). Note the change of 1,500Wh battery configuration for spacecraft 16. This configuration also applied to spacecraft 19, which never flew. *(NASA)*

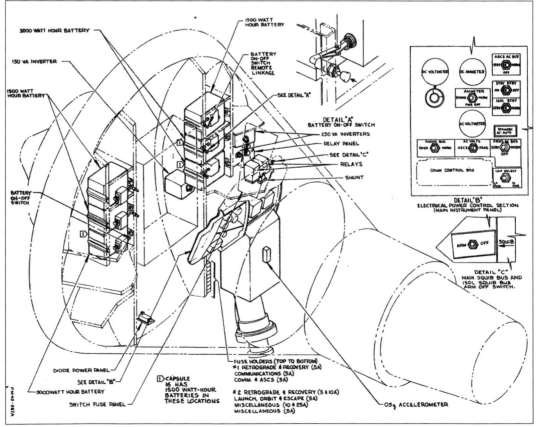

DC power

Three 3,000Wh batteries, each weighing 45lb (20.4kg), provided the primary 24V direct current power supply and were connected in parallel to the circuit via an on-off switch on the battery, reverse current diodes providing protection from discharge through a faulty battery or one very low on power. The main DC bus was connected directly to the 250V and 150VA (volt ampere) inverter filter inputs. Separately, two 1,500Wh stand by batteries were carried for communications bus voltage and to serve as standby for the primary power distribution subsystem. These incorporated reverse current protection and like the primaries were connected to the circuit through on-off switches on each battery.

Taps on the standby batteries provided 6, 12 and 18V DC through fuses to the various system and subsystem busses and before launch these were energised through external power loops via the ground umbilical disconnect and external power fuses to protect the spacecraft circuits. The 24V DC power from the standby batteries

was controlled by the 'STDBY BATT' switch mounted on the main instrument panel, which could be placed in either the on or off position by the astronaut. When in the off position the main batteries supplied 24V DC power to the spacecraft electrical systems through the main bus. In the on position the DC bus was connected to the main bus and both battery groups provided electrical power in parallel.

If a launch hold was called after the ground umbilical was disconnected a 'HOLD' command would be initiated to activate an emergency circuit and maintain power. The emergency hold circuit removed power from non-essential equipment but allowed visual observation through the periscope. It also applied power to the cabin vent squibs and to the periscope motor for effective operation by the astronaut. In that event the hold signal from the rocket would be directed through circuits which would normally be closed but power passed through the ground test umbilical relay to the solenoid of the No 1 emergency hold relay.

Power from the main DC bus would then be carried to the solenoid of No 2 and No 3

BELOW With the proviso that specific configurations of battery capability changed on a per-spacecraft basis, this typical arrangement provides a clear indication of power routes and various systems for both AC and DC circuits. It also indicates the external power connections where relevant.
(McDonnell)

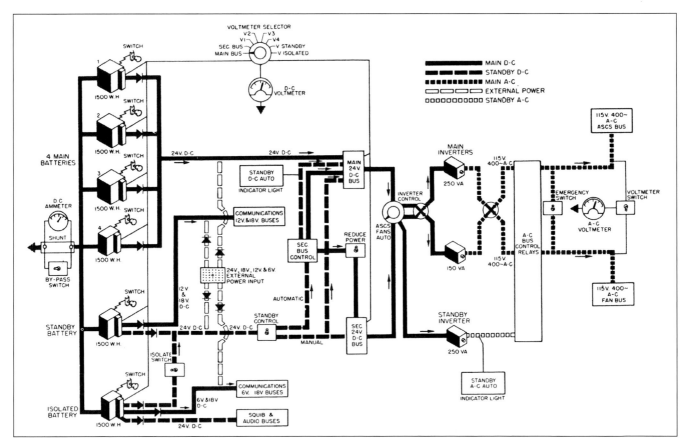

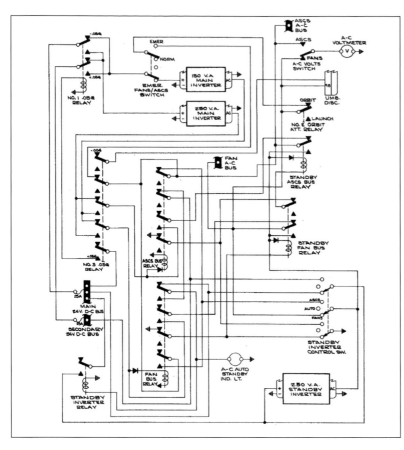

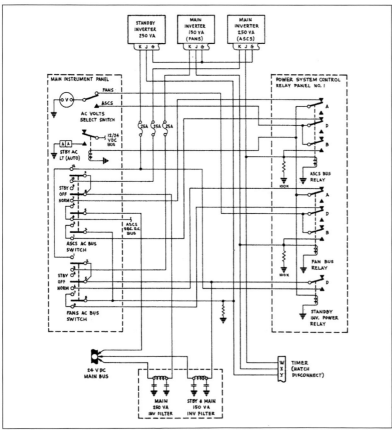

emergency hold relays. By actuating these, power would be removed from the non-essential circuits and the impact sensor relay would be energised. Simultaneously, power would be carried through the normally open contacts of the emergency hold relays to extend the periscope and fire the cabin vent squibs.

Separately, a 1,500Wh isolated battery provided emergency audio bus and squib-firing voltages and also provided emergency power to other circuits if the main and standby batteries were exhausted. It also provided reverse current protection and was connected to the isolated bus through the on-off switch. The battery taps supplied 6V and 18V DC power to selected system busses through self-contained diodes. The 24V isolated battery output was available to associated busses through the 'ARM' position of the squib arming switch through the 'EMERG' position of the 'AUDIO BUS' switch. Power from the isolated battery could also be connected in parallel with the 24V output of the standby batteries by placing the 'ISOL-BTRY' switch in the 'STDBY' position.

Direct current electrical power was provided to the spacecraft on the launch pad through the umbilical cable, which was used for pre-launch operations to conserve the onboard battery life. It was normal for the 24V DC supply to be delivered to three inputs through three fuses in the spacecraft, whence it was directed to appropriate busses. Four external power supplies were used to supply 6, 12, 18 and 24V as required, with individual battery and main bus voltages read by the astronaut using a voltmeter and selector switch. The direct current ammeter was used, with the switch in the 'NORM' position, to read out these individual battery and main bus voltages. A Zener diode panel was provided adjacent to the Zener diodes to connect them to the main and isolated 24V DC busses to provide protection from transient spike voltages affecting the transistorised equipment.

AC power

Two inverters supplied alternating current for the main 115V, 400-cycle, alternating current power supply of 150VA and 250VA each. This AC load was divided into two groups, the ASCS bus and the FANS bus. The 250VA inverter provided power to the ASCS bus and the 150VA to the FANS. The main DC bus powered the 150VA inverter through a line filter circuit and a 24A fuse. The 250VA inverter was similarly powered through the main DC bus via a line filter and 25A fuse. The DC power was controlled via the 'NORM' position of the 'ASCS', 'AC BUS', and 'FANS AC BUS' switches on the main instrument panel in the capsule. The outputs of the two inverters fed solenoids on the two bus relays and these energised the inverter output through closed contacts of the relays powering the FANS and ASCS busses.

AC power from the standby inverter was at 115V, 400-cycle, supplied by one 250V inverter that would deliver energy to either the ASCS, the FANS or both AC busses upon selection of the 'STDBY' position on the appropriate 'ASCS AC BUS' or 'FANS AC BUS' switch found on the main instrument panel.

If either of the 250VA or 150VA main inverters should fail the respective FANS bus relay of the ASCS bus relay would be disabled, and that would simultaneously de-energise the standby inverter relay, which would in turn apply DC power from the filter to that inverter. The AC output from the standby inverter would then be directed through contacts in the de-energised ASCS or FANS bus relay to their respective busses. The astronaut would have received a warning through an indicator on the main instrument panel that the standby had automatically switched on due to a failure of either of the main inverters.

The DC-AC inverters were of solid-state design and could operate continuously at full rated power in an ambient temperature of 160°F (71.1°C) or 80°F (26.7°C) at 5lb/in² (34.47kPa) in a 100% oxygen environment. Output was 115V AC, ±5%, single phase to ground at 400 cycles, ±1%, which were essentially sinusoidal in waveform. Qualification tests verified that a short circuit across the inverter output side would not damage the inverter itself or the wiring involved.

DC power distribution

Direct current was taken from the three separate battery groups (main, standby and isolated) described above and various

BELOW The functional DC power control electrical system for early spacecraft is displayed on this schematic that delineates standby circuits and battery configurations. *(McDonnell)*

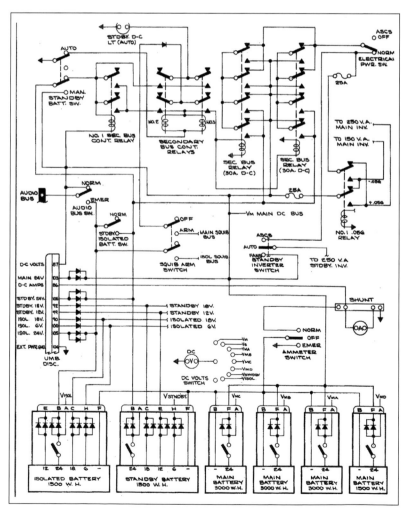

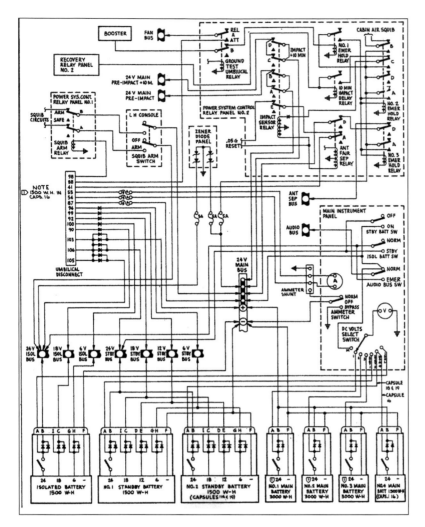

ABOVE The DC power circuits for the advanced Mercury 16, 18 and 19 spacecraft reflect changes in the battery types and capacities compared to the earlier configurations, principally with the addition of a second 1,500Wh standby battery. *(McDonnell)*

sub-busses operating from these three power sources. The bus separation methods were as follows: (a) main AC bus direct to the main batteries; (b) main 24V DC antenna separation bus from main bus through separation relay; (c) pre-impact plus ten minutes from main bus through impact relay; (d) main 24V squib bus through 'SQUIB ARM SW' from the main bus; (e) pre-impact main bus through impact relay from the main bus; (f) audio bus from main bus to isolated bus through 'AUDIO BUS SW'; (g) isolated DC bus directly to isolated battery; (h) standby DC bus direct to standby battery; (i) isolated 6V and 18V busses direct to taps on isolated

battery; and (j) standby 6, 12 and 18V busses direct to taps on standby battery.

Batteries

The batteries consisted of a series of connected silver-zinc rechargeable cells with a nominal rating of 24.5V and a minimum capacity rating of 3,000Wh for the three main batteries, 1,500Wh for each of the two standby batteries and the isolated battery. Each of the six batteries supported an externally mounted relief valve adjusted to maintain an internal pressure range of between 5.5lb/in^2 (37.9kPa) and 14.9lb/in^2 (102.7kPa). The rated capacity of the batteries was 40Ah, but brief pulsed current could rise to 42A.

Each battery had an electrolyte solution of 40% reagent-grade potassium hydroxide and distilled water and was used to activate the dry charge immediately before use. Following the first discharge cycle the battery could be recharged by any battery charger delivering a constant current. Great emphasis was placed on the preparation and servicing of the batteries, since they were the sole means of electrical power aboard, without which the spacecraft would be unable to return the astronaut safely to Earth.

The Eagle-Picher batteries in the Mercury spacecraft were designed and qualified for five complete cycles of discharge and charge but there was a desire not to conduct more than four cycles or an activated life of 60 days before flight.

The astronaut had two key instruments on the main display panel for determining the state of the batteries and the health of the electrical distribution system: the DC ammeter and the DC voltmeter. The ammeter carried information on total current drain from the six batteries and had a basic movement sensitivity of 50mV. A suitable resistance was connected across the input of the meter to provide a low resistance path to ground with the proper voltage drop at 50A for a meter movement at full-scale deflection.

The DC voltmeter was located on the main panel with a switch with which the astronaut could select a specific battery for reading the individual voltages on a scale of 0–30V.

Lighting

The interior of the Mercury spacecraft was illuminated naturally through the observation window and the periscope, and with a 90-minute orbit and a sunrise on every one there were lighting considerations that ranged from brilliant sunlight to deep blackness. To counter the optical stress of rapidly changing light levels Mercury could isolate its pilot with shades, but a general requirement for balanced illumination inside provided the pilot with four fluorescent floodlights and a series of waning lights adjusted in intensity according to the ambient levels.

The cabin floodlights were mounted on brackets to the left and right and above the astronaut's head. Power was provided from the 115V AC bus and it was controlled by a three-position switch located on the left side console. The switch positions were marked 'BOTH', 'LH ONLY' and 'OFF'. The cabin lights had high actinic value, which was particularly suitable for camera usage in space. The lights gave out very little heat and were rated at 4W each.

Two fluorescent photo floodlights were identical to the cabin floodlights and were mounted on the left instrument console, controlled by a 'PHOTO LIGHTS' switch with a single 'ON' and 'OFF'. Various consoles had warning telelights that were mounted on the main instrument panel and on the left console. These were connected to various systems to alert the astronaut to a malfunction of a particular system or simply to notify him of a specific function. Power for these telelights came from the capsule's 12V DC or 24V DC bus via a 5A fuse. To accommodate the changing day or night light levels from outside a 'DIM-BRIGHT' switch was provided.

Conforming to the reality that no two spacecraft were the same, the precise configuration of specific telelight annotations varied mission by mission and according to the test objectives of each flight. The details presented here are typical of a representative flight configuration, but readers should be aware that individual spacecraft carried a wide range of minor changes to the configuration of notifications via these telelights. The configuration of the bus circuits also varied to a minor degree.

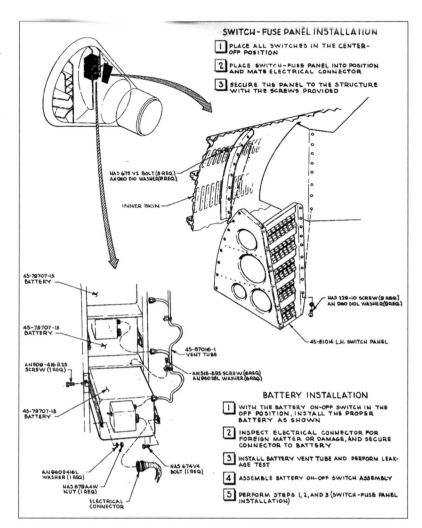

ABOVE Taken from the pages of the maintenance manual, this shows the procedures for installing batteries and fuse panel located on the left side of the pressure cabin. (McDonnell)

BELOW Power demand fluctuated according to the activated systems and their specific application. This power train timeline for the 15-minute flight of Gus Grissom on MR-4 shows a typical consumption level for AC, DC and current in amps, logged by sequential event. (NASA)

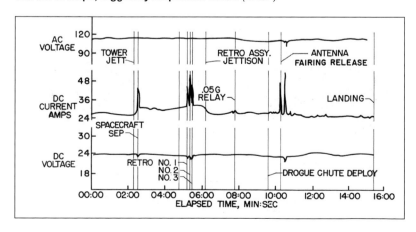

Environmental Control System (ECS)

The primary purpose of the ECS was to support the life of the pilot and to maintain a physiologically stable environment, controlling the gas composition, pressure and temperature in a pressurised cabin. It was also designed to support a space suit, for backup in the event of a failure to the cabin pressurisation system.

The entire system was developed and manufactured by the McDonnell Aircraft Company following a contract awarded in February 1959. But MAC sought the advice of AiResearch, a division of the Garrett Corporation, and before the contract was awarded the company met with them and personnel from the STG to decide on the design criteria. AiResearch was subcontracted to MAC to provide the ECS. But there were fundamental decisions pending regarding the type of environment for Mercury and, as it turned out, several future generations of spacecraft beyond.

In 1959 physiologists were uncertain as to how long a human could breathe pure oxygen without adverse pulmonary or vascular consequences, which at the time were largely unknown. Challenged by aircraft flying faster and higher into realms where the human body required special protection, the Air Force and the Navy carried out extensive experimental flights and simulations on the ground to study these effects.

The Earth's atmosphere is 21% oxygen and 78% nitrogen, the balance containing a mixture of several different gases including carbon dioxide, helium, hydrogen, neon and krypton. While the human body can exist for some time without nitrogen it needs oxygen to perform, but the human brain is an oxygen-hungry organ; 25% of the energy required to live is taken up by the brain (which is why physical work is tiring but mental work is exhausting).

Sea-level atmospheric pressure is $14.7lb/in^2$ (101.35kPa) and it is possible for people to survive and work for a limited period with undiminished performance if all the nitrogen is removed and the base level of 3.53lb/in (24.34kPa) oxygen is maintained. Below that level oxygen starvation begins to set in, causing disorientation, dysfunction of motor skills, sleepiness and death. But physiologists were uncertain about the length of time humans could endure that oxygen-rich environment. Moreover, there were other, dangerous, consequences.

Engineers were aware that in a fire the flame propagation rate is almost exponentially proportional to the increased intensity of the oxygen, from 21% to 100%. Very few expressed concern that a pure oxygen environment could have disastrous consequences if a fire broke out inside the pressurised capsule in space. However, the nuances of weightlessness had not quite hooked up in terms of defining how flames behave in weightlessness but the physics told them that, at least in principle, where there is no convection the intensity of the flame would not be as great as in a 1g environment where heat rises. Without convection the flame would quickly smother itself, so there was relief from the fear that a pure oxygen environment was deadly should a fire break out.

The preference for a pure-oxygen design was driven to some extent by the fact that a two-gas system was complicated, more difficult to regulate and harder to engineer with

RIGHT The Mercury environmental control system was the first engineered for a US manned space vehicle and provided pressurisation, temperature and humidity control, and removal of exhaled carbon dioxide for the astronaut. It was designed to support the astronaut in a pressure suit in the event of depressurisation. It also supported an active thermal control system by means of a coolant supply. (NASA)

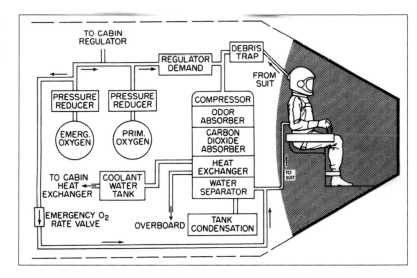

appropriate backup and redundancy. This was due to the increased complexity and multiplicity of components and subsystems required to make it work with the same reliability and safety criteria. Moreover, by reducing the internal pressure to 21% of a two-gas system, the physical structure of the pressure vessel would be lighter and allow a greater margin of safety.

The decision was driven by the STG, who saw no credible engineering or physiological reason to adopt the more complex two-gas system when every combat aircraft in the world operated on a pure oxygen environment. Max Faget in particular was averse to going against this proven record of minimum systems complexity and maximum operational success. As such, the Mercury spacecraft would provide a habitable environment through the dual provision of a single-gas system with a pressurised space suit as backup, and only in the suit would it challenge state-of-the-art engineering and technology of the day. McDonnell had considerable experience with aircraft ECS designs for a variety of first-generation jet fighters built for the Navy. Nevertheless, providing a man in space with a habitable environment, controlled and with a high degree of reliability, was a different matter.

With agreement for a single-gas system for pressurisation and for breathing, and the provision of a pressurised suit in the event of depressurisation of the cabin, heat rejection was calculated as 400Btu/hr (422kj/hr) with an oxygen consumption of 30.5in³/min (500cm³/min) at standard temperature and pressure. Carbon dioxide production was calculated at 2.6lb/day (1.179kg/day) and perspiration and water at 6lb/day (2.72kg/day). The cabin temperature was to be maintained at 80°–100°F (26.7°–37.8°C). Purging the cabin of an oxygen/nitrogen sea-level mix was to be conducted by using a separate oxygen supply with reducers.

Development of an effective ECS progressed rapidly through 1959, with a test chamber ready by September, but difficulties with detailed design held up progress. The STG went to the

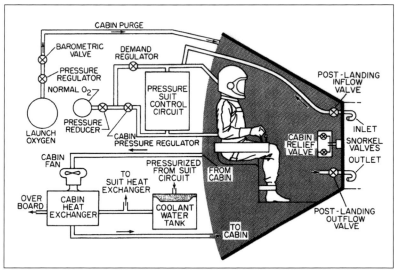

Duke University School of Medicine to recruit physiologists sufficiently experienced to help with giving the technology a man-rating label. A major problem lay in the lack of information about how a human would respond to the conditions in space, with weightlessness and significant fluctuations in external temperatures. A significant challenge was in the requirement for a system capable of performing as well after 28 hours as it was at launch.

Mock-ups and manikins were used to test out the system but progress was slow and dragged on. Not until early 1961 was the ECS ready for certification. AiResearch had conducted many on-site hardware evaluations during the development cycle. The pioneering nature of the subsystems only served to emphasise the groundbreaking work that would form the base technology from which both Gemini and

ABOVE The ECS was designed to ensure a habitable environment in the spacecraft and to include the suited astronaut in that loop for contingency or emergencies. It is to the credit of the engineering design that the suit was never used, all life-support functions being supported by the system displayed on this schematic. (McDonnell)

RIGHT The ECS functional components displayed in 'breadboard' fashion and not in the manner in which they were installed in the spacecraft. *(McDonnell)*

BELOW Taking the functional components in the breadboard schematic, this illustration displays the way in which the various ECS elements were installed. Specific spacecraft varied in some degree but this represents the baseline used for all manned vehicles. *(McDonnell)*

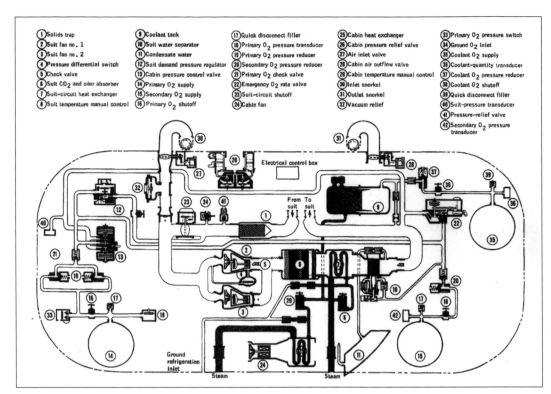

① Solids trap	⑨ Coolant tank	⑰ Quick disconnect filler	㉕ Cabin heat exchanger	㉝ Primary O_2 pressure switch
② Suit fan no. 1	⑩ Suit water separator	⑱ Primary O_2 pressure transducer	㉖ Cabin pressure relief valve	㉞ Ground O_2 inlet
③ Suit fan no. 2	⑪ Condensate water	⑲ Primary O_2 pressure reducer	㉗ Air inlet valve	㉟ Coolant O_2 supply
④ Pressure differential switch	⑫ Suit demand pressure regulator	⑳ Secondary O_2 pressure reducer	㉘ Cabin air outflow valve	㊱ Coolant-quantity transducer
⑤ Check valve	⑬ Cabin pressure control valve	㉑ Primary O_2 check valve	㉙ Cabin temperature manual control	㊲ Coolant O_2 pressure reducer
⑥ Suit CO_2 and odor absorber	⑭ Primary O_2 supply	㉒ Emergency O_2 rate valve	㉚ Inlet snorkel	㊳ Coolant O_2 shutoff
⑦ Suit-circuit heat exchanger	⑮ Secondary O_2 supply	㉓ Suit-circuit shutoff	㉛ Outlet snorkel	㊴ Quick disconnect filler
⑧ Suit temperature manual control	⑯ Primary O_2 shutoff	㉔ Cabin fan	㉜ Vacuum relief	㊵ Suit-pressure transducer
				㊶ Pressure-relief valve
				㊷ Secondary O_2 pressure transducer

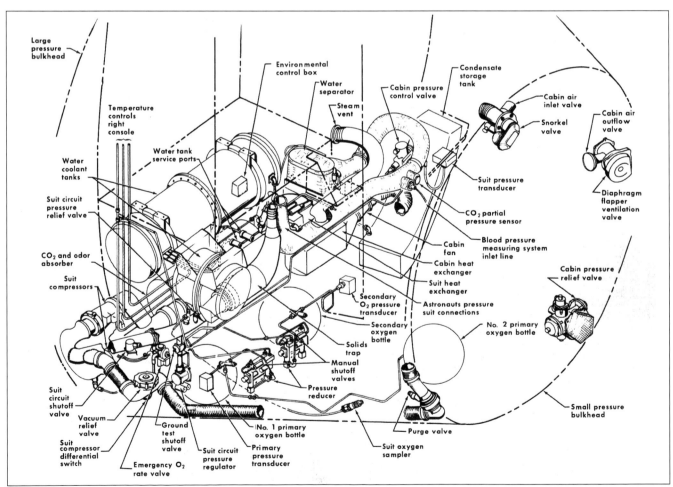

Apollo would emerge. Challenging problems involved with countering hypoxia, dysbarism and hyperventilation were to keep ECS engineers and test subjects busy for two years.

The design for the Mercury spacecraft is known as a semi-closed system where carbon dioxide (CO_2) is removed from the atmosphere by chemical absorption, where water vapour is removed by condensation and stored, and oxygen is recirculated by a fan. Two 7,500lb/in²g (51,712kPag) storage vessels provided the oxygen and metabolic heat, and thermal energy from equipment was rejected through evaporation of water at about 35°F (1.67°C) through heat exchangers, using water from a storage tank delivered through positive expulsion.

The circuit for the pressure space suit provided a 34in (86cm) long pipe that carried gas removed from the helmet to a solids trap fabricated from stainless steel that filtered out particles larger than 40 microns. The smoothbore hose was made of silicone rubber with an internal diameter of 1in (2.5cm). A bypass, provided in case the filter became clogged, was activated when the differential pressure across the trap increased to 0.5in (1.25cm) of water. On leaving the trap the gas passed to the inlet of the suit compressor, which then recirculated oxygen through to the suit circuit at a rate of 11.4ft³/min (0.3m³/min).

The centrifugal compressor which recirculated the oxygen was driven by a 115V, 400-cycle, single-phase electric motor operating through a step-up gear with a ratio of about 1.8:1, producing an impeller speed of 43,000rpm. The compressor operated with a differential water pressure of 10in (25.4cm) of water when recirculating at a density of 0.0272lb/ft³ (0.4357kg/m³). Redundancy was provided by a standby compressor mounted parallel to the primary, with automatic or manual activation. A differential switch automatically switched from primary to secondary should the suit circuit drop below 10in of water, but only when the suit-fan selector was in the normal position. Separate switch positions allowed the pilot to move from automatic to manual control.

A common discharge duct from the two compressors directed the flow to the inlet of the CO_2 and the odour absorber via a flexible bellows that allowed replacement and provided

isolation from vibrations in the motor. Internally, the absorber bifurcated into two identical paths. In linear progression, each contained a 0.5lb (0.23kg) block of activated charcoal and two 1.15lb (0.52kg) charges of 4–8 mesh lithium hydroxide (LiOH), or a total 1lb (0.45kg) of activated charcoal and 4.6lb (2.08kg) of LiOH. The activated charcoal removed odours from the closed suit circuit while the LiOH absorbed CO_2. This operated on the chemical reaction $2\ LiOH + CO_2 \rightarrow Li_2CO_3 + H_2O$ + heat. The six charges were contained in a single fibreglass-covered cloth bag.

Gas leaving the absorber passed through an Orlan filter with the charge compressed by spring so as to inhibit separation of the LiOH during launch vibrations. Separation could have caused channelling of the flow, which would briefly bypass the absorber and compromise removal of carbon dioxide. After emerging from the absorber the gas was directed to the suit heat exchanger which cooled it to

BELOW Key elements of the ECS are displayed here with electrical switch positions together with console readouts and actuating switches on the main instrument console. *(NASA)*

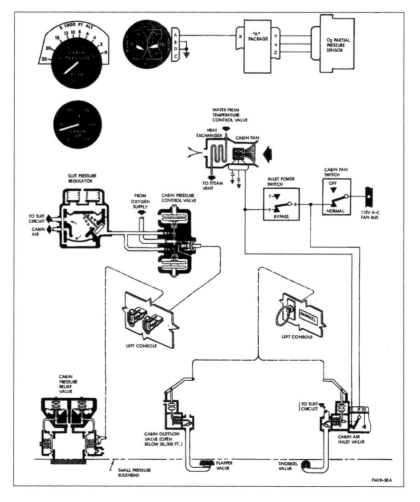

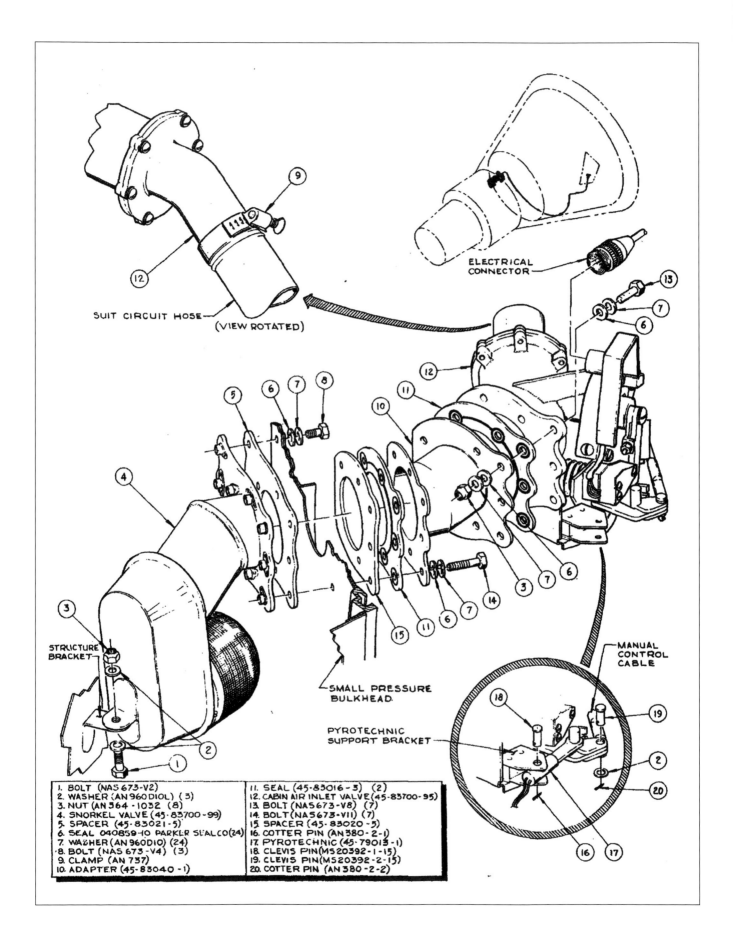

SUIT CIRCUIT HOSE

(VIEW ROTATED)

ELECTRICAL CONNECTOR

STRUCTURE BRACKET

SMALL PRESSURE BULKHEAD

MANUAL CONTROL CABLE

PYROTECHNIC SUPPORT BRACKET

1. BOLT (NAS 673-V2)
2. WASHER (AN 960 DIOL) (3)
3. NUT (AN 564-1032) (8)
4. SNORKEL VALVE (45-83700-99)
5. SPACER (45-83021-5)
6. SEAL 040859-10 PARKER SEALCO (24)
7. WASHER (AN 960 DIO) (24)
8. BOLT (NAS 673-V4) (3)
9. CLAMP (AN 737)
10. ADAPTER (45-83040-1)

11. SEAL (45-83016-3) (2)
12. CABIN AIR INLET VALVE (45-83700-95)
13. BOLT (NAS 673-V8) (7)
14. BOLT (NAS 673-VII) (7)
15. SPACER (45-83020-5)
16. COTTER PIN (AN 380-2-1)
17. PYROTECHNIC (45-79013-1)
18. CLEVIS PIN (MS 20392-1-15)
19. CLEVIS PIN (MS 20392-2-15)
20. COTTER PIN (AN 380-2-2)

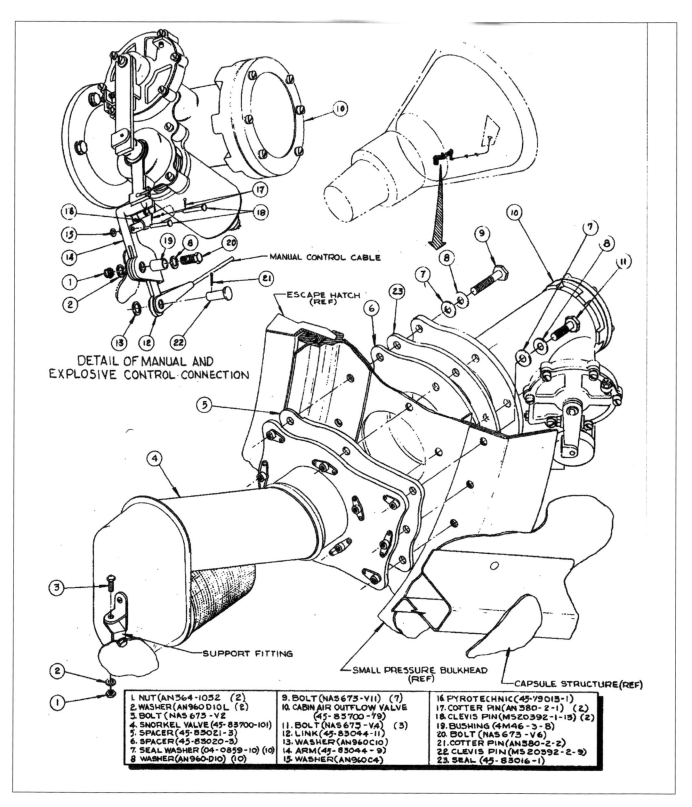

DETAIL OF MANUAL AND
EXPLOSIVE CONTROL CONNECTION

MANUAL CONTROL CABLE

ESCAPE HATCH
(REF)

SUPPORT FITTING

SMALL PRESSURE BULKHEAD
(REF)

CAPSULE STRUCTURE (REF)

1. NUT (AN364-1052 (2)	9. BOLT (NAS675-V11) (7)	16. PYROTECHNIC (45-79013-1)
2. WASHER (AN960D10L (2)	10. CABIN AIR OUTFLOW VALVE	17. COTTER PIN (AN580-2-1) (2)
3. BOLT (NAS 673-V2	(45-83700-79)	18. CLEVIS PIN (MS20392-1-13) (2)
4. SNORKEL VALVE (45-83700-101)	11. BOLT (NAS675-V4) (3)	19. BUSHING (4M46-3-8)
5. SPACER (45-83021-3)	12. LINK (45-83044-11)	20. BOLT (NAS 673-V6)
6. SPACER (45-83020-3)	13. WASHER (AN960C10)	21. COTTER PIN (AN580-2-2)
7. SEAL WASHER (04-0859-10) (10)	14. ARM (45-83044-9)	22. CLEVIS PIN (MS20392-2-9)
8. WASHER (AN960-D10) (10)	15. WASHER (AN960C4)	23. SEAL (45-83016-1)

OPPOSITE The cabin air inlet valve ad snorkel valve installation allowed the pressurised cabin to receive air from the atmosphere for equalisation and also for venting during ascent. *(McDonnell)*

ABOVE The cabin air outflow valve and snorkel valve installation showing manual and explosive control interconnection, the latter as a precaution against over pressurisation. *(McDonnell)*

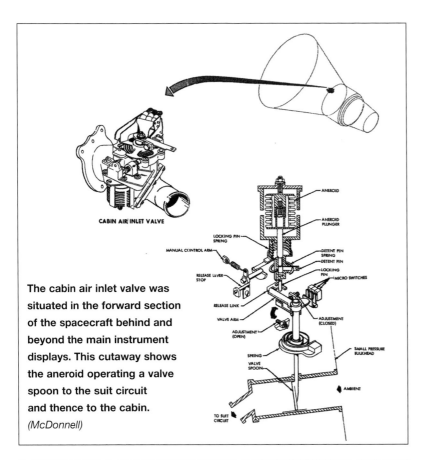

The cabin air inlet valve was situated in the forward section of the spacecraft behind and beyond the main instrument displays. This cutaway shows the aneroid operating a valve spoon to the suit circuit and thence to the cabin. *(McDonnell)*

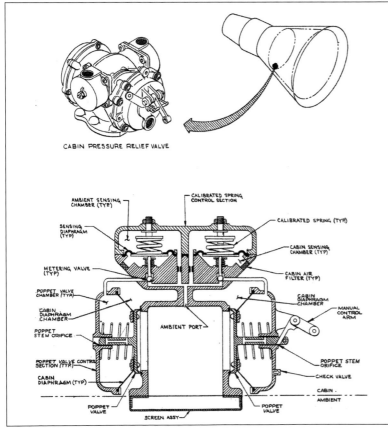

45°F (7.2°C), at which temperature metabolic water and water from the LiOH-CO_2 reactions were condensed. The heat rejected to water evaporates in the vacuum of space at about 35°F (1.7°C), which under such condition was about 1,075Btu (1,133kj/hr) of energy. Total heat load in the suit system was about 700Btu/hr (737.8kj/hr) including the heat from the astronaut, the heat from the reaction of lithium hydroxide and the heat from the compressor motor. This total load was equal to a water flow rate of 0.7lb/hr (0.3kg/hr).

The heat exchanger itself was made of aluminium plate finned to maximise thermal emissivity with alternate passages for gas and evaporating water. Heat transfer was of the crossflow type where the gas makes a single mass and the evaporating water executes two passes. A needle valve operated by the astronaut allowed manual adjustment of the water flow. Water for the suit and cabin heat exchanges was stored in a coolant tank with a capacity of 39lb (17.7kg). This contained a bladder for positive expulsion in weightless conditions.

The water that was condensed in the heat exchanger was collected by the suit water separator, a sponge squeezed by a pneumatic piston. The spacecraft programming unit was programmed to squeeze the sponge for 30 seconds every 30 minutes by using a pressure of 100lb/in² (689.5kPa), the liberated water being discharged through a 5lb/in²g (34.47kPag) check valve into a condensate tank wherein a sintered bronze plug released gas to the cabin without the condensate.

Cleansed of carbon dioxide and odours and cooled and dehumidified, the recirculated oxygen was returned to the suit inlet fitting situated at the waist. Four separate flexible tubes ducted gas to the hands and feet with a configuration that allowed oxygen to flow over the body. In this process it collected water vapour inside the suit and CO_2 from the helmet area, the gas then exiting via the helmet fitting.

LEFT A bulky device and surprisingly heavy, the cabin pressure relief valve was located adjacent to the cabin air inlet valve and could be overridden by a manual handle that was easier to reach than the unit itself. *(McDonnell)*

RIGHT The configuration of the primary and secondary oxygen supply loops with service and manual shut-off valves and associated electrical actuation switches as well as warning lights – which could be dimmed – displayed on the instrument panel. Note that the pressure transducers are slaved to instrumentation package 'A'. *(McDonnell)*

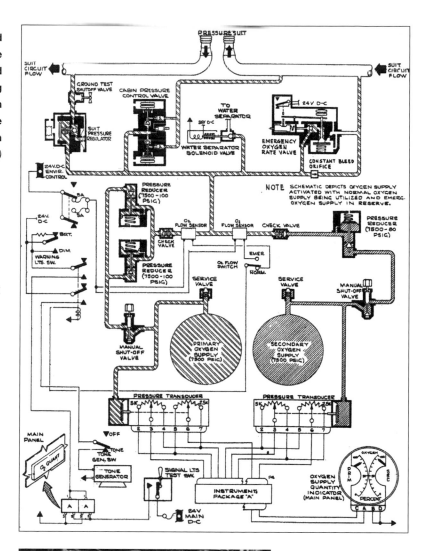

Oxygen make-up for losses in the metabolic process was accomplished by one of two methods. In one, the suit demand regulator provided oxygen to the suit circuit when pressure fell to 3.0in (7.62cm) of water below cabin pressure, thereby allowing a small amount of oxygen to increase suit pressure above this level. This demand regulator was connected to the suit circuit downstream of the solids trap but a relief diaphragm was provided when suit pressure exceeded cabin pressure by 2in (5cm) of water above cabin pressure. Both demand and relief systems would sense the pressure of the cabin but the regulator would automatically provide a reference pressure of 4.6lb/in²g (31.7kPag), ±0.2lb/in²g (1.38kPag) if cabin pressure fell below that level.

Artificial reference pressure was provided by either of two redundant aneroids and also by a constant-bleed orifice. The aneroids were designed to close when pressure in the cabin fell below 4.6lb/in²g (31.58kPag), which also isolated the demand and relief diaphragm reference chambers from cabin pressure. The constant-bleed orifices had a 1.83in³/min (30cc/min) maximum flow rate at standard temperature and pressure and would maintain pressure within the reference chambers, providing case leakage of the demand regulator.

The second method for oxygen supply make-up was through operation of the cabin pressure control valve, which sensed pressure and directed flow through the suit circuit, maintaining a pressure level slightly higher than the cabin and thus preventing contaminants flowing into the suit itself. It operated by reacting to a cabin pressure drop below 5.1lb/in²g (35.16kPag) and supplied an oxygen flow of 0.010lb/min (0.0045kg/min) to compensate for the suspected leakage. It continued to flow at this rate until it exceeded the relief diaphragm

LEFT Gordon Cooper about to ride a centrifuge, seated in its gondola at the Naval Air Development Center, Warminster, Pennsylvania. The NADC supported high-g tests and rigorous evaluation, some data from which was crucial to developing the Mercury ECS by measuring the astronauts and their oxygen uptake and exhalation requirements. *(US Navy)*

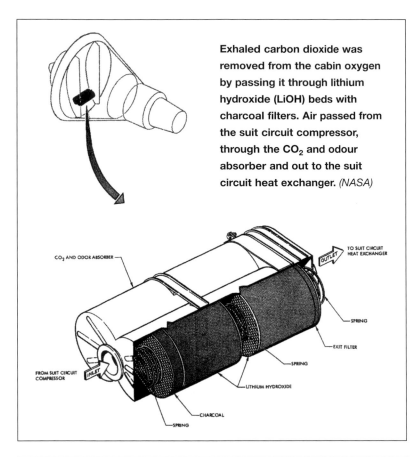

Exhaled carbon dioxide was removed from the cabin oxygen by passing it through lithium hydroxide (LiOH) beds with charcoal filters. Air passed from the suit circuit compressor, through the CO₂ and odour absorber and out to the suit circuit heat exchanger. *(NASA)*

CO₂ AND ODOR ABSORBER

OUTLET — TO SUIT CIRCUIT HEAT EXCHANGER

SPRING

EXIT FILTER

FROM SUIT CIRCUIT COMPRESSOR — INLET

SPRING

LITHIUM HYDROXIDE

CHARCOAL

SPRING

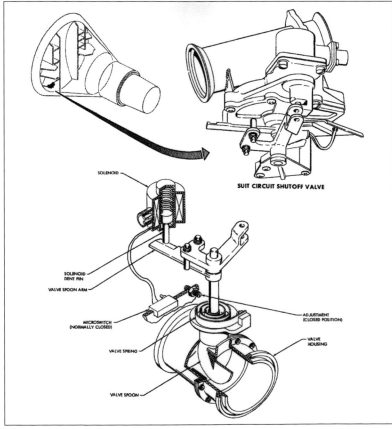

SUIT CIRCUIT SHUTOFF VALVE

SOLENOID

SOLENOID DENT PIN

VALVE SPOON ARM

MICROSWITCH (NORMALLY CLOSED)

VALVE SPRING

VALVE SPOON

ADJUSTMENT (CLOSED POSITION)

VALVE HOUSING

in the demand regulator that relieved it when the differential pressure had reached 2in (5cm) of water. It continued in this mode until back to the 5.1lb/in² level, when it turned off the valve.

If cabin pressure leakage exceeded 0.010lb/min it would decay down to zero, but the suit pressure make-up would continue to supply oxygen at this rate. The suit would stop deflating at a pressure of 4.6lb/in²g (31.7kPag) and the cabin pressure relief valve would stop supplying oxygen to the cabin if pressure fell below 4.0lb/in²g (27.58kPag), +0.2/-0.1lb/in²g (1.38/0.69kPag). The cabin pressure relief valve had redundant control aneroids and metering orifices for contingency and the valve had a cabin repressurisation capability of 0.17lb/min (0.77kg/min), but this was manually operated and independent of the cabin pressure.

The gas supply consisted of two 4lb (1.8kg) tanks of gaseous oxygen stored at 7,100lb/in²g (48,954kPag), separately designated as primary and secondary supply sources. Providing the leak rate was less than 18.3in³/min (300cm³/min) each tank alone was capable of supporting a 28-hour mission. Each tank had an internal volume of 198in³ (3,245cm³) and was fabricated from 4340 carbon steel with nickel plating resistant to corrosion. The tanks had a specified proof pressure of 12,500lb/in²g (86,187kPag) and a burst pressure of 16,700lb/in²g (115,146kPag) and each had a high-pressure shut-off valve, a filler valve and a pressure transducer required to measure the quantity of oxygen remaining.

A single-stage pressure reducer lowered primary tank pressure to 100lb/in² (689.5kPa), ±10lb/in² (68.95kPa), and two were provided for redundancy. The secondary tank had a single pressure reducer that regulated the supply to 80lb/in²g (552kPag). Both primary and secondary tanks were connected to a manifold downstream of their reducers, configured so that when the primary tank was depleted the pressure in the manifold linking the two systems would decay to 80lb/in², allowing the flow to

LEFT Located at the forebody end of the spacecraft and to the lower right of the couch position, the suit circuit shut-off valve operated via a valve spoon arm and solenoid. *(McDonnell)*

begin automatically from the secondary tank. In the event of a failure in either system, reverse flow was inhibited by a check valve at the discharge end of each reducer.

The common manifold was usually operated at a pressure of $100lb/in^2g$ (689.5kPag), $\pm10lb/in^2$ (68.95kPa), providing oxygen to the four ECS valves, three of which were the suit pressure demand regulator, the cabin pressure control valve and the water separator solenoid. An emergency oxygen-rate valve provided the suit circuit with automatic activation when suit pressure fell below $4.0lb/in^2g$ (27.58kPag) but it could also be manually actuated via an 'EMER O_2' handle situated on the right side of the pressurised cabin. An aneroid installed within the emergency oxygen rate valve maintained continuous monitoring of decreasing suit pressure and would energise a solenoid in the shut-off valve, located downstream of the solids trap, consisting of a spoon valve spring-loaded in the closed position but held open by a pin integral with the core of the solenoid.

When the microswitch in the emergency rate valves closed it energised the solenoid, retracting the pin. As the spring closed the shut-off valve rotated a shaft in the rate valve, uncovering an orifice and starting an oxygen flow rate of 0.049–0.075lb/min (0.022–0.034kg/min). The mechanical linkage extended from the rate valve to the right-hand console, moving the 'EMER O_2' handle to the emergency position. At this point a suit fan relay would interrupt power to the operating suit compressor and the check valve located downstream of the water separator would close. Isolating the compressors, the CO_2 and odour absorber, the heat exchanger and water separator from the pressure suit, the emergency flow rate would be relieved to the pressure cabin via the relief port of the suit pressure regulator.

At the time the suit pressure increased to $4.0lb/in^2$ the pilot could return the handle to the normal position, which automatically returned the system to a standard mode of operation. In completing this action the suit-fan cut-off relay would be de-energised and the suit compressor operated, the system shut-off valve would open and latch in that position and flow from the emergency-rate valve would be interrupted. This emergency mode could be operated by the

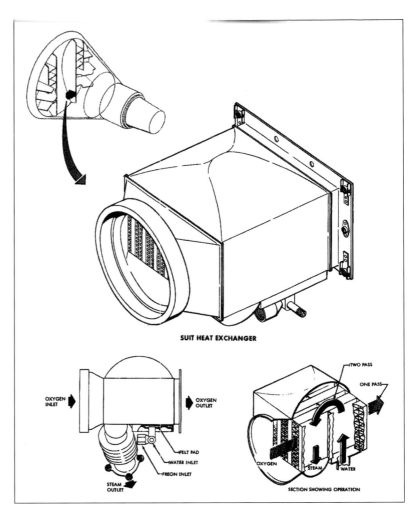

ABOVE The suit heat exchanger accepted oxygen and took water to produce steam, delivered via a regulator, for controlling temperature and humidity. *(McDonnell)*

BELOW The suit pressure regulator accepted oxygen from the cabin circuit and fed through a poppet valve to a screen and outlets to the cabin via the relief port. *(McDonnell)*

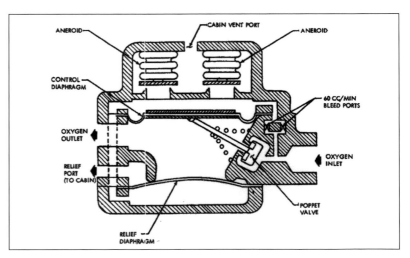

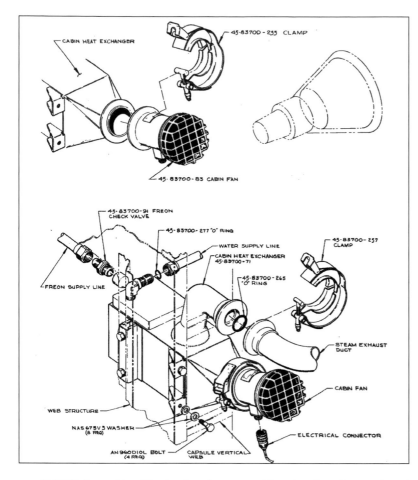

Labels in diagram:
CABIN HEAT EXCHANGER
45-83700-255 CLAMP
45-83700-83 CABIN FAN
45-83700-91 FREON CHECK VALVE
45-83700-277 'O' RING
WATER SUPPLY LINE
CABIN HEAT EXCHANGER 45-83700-71
45-83700-257 CLAMP
45-83700-265 'O' RING
FREON SUPPLY LINE
STEAM EXHAUST DUCT
CABIN FAN
WEB STRUCTURE
NAS 673V3 WASHER (8 REQ)
AN 960D10L BOLT (4 REQ)
CAPSULE VERTICAL WEB
ELECTRICAL CONNECTOR

ABOVE Assembly of the heat exchanger and fan showing steam exhaust duct and electrical connector. *(McDonnell)*

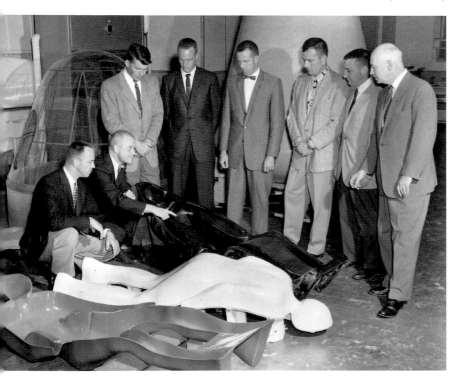

astronaut at any time by moving the 'EMER O$_2$' handle to the emergency position; this would rotate the shaft of the rate valve which, having an integral helical ramp, lifted the pin which thereby closed the microswitch.

In addition to the cabin pressure control valve described earlier, the ECS subsystems for the pressurised cabin included the cabin fan and associated heat exchanger, the cabin pressure relief valve, the cabin air inlet valve and the cabin air outflow valve. The fan delivered oxygen to the heat exchanger, where heat generated by the electrical equipment was removed. A vane-axial design driven by a 115V, 400-cycle, single-phase electric motor rotated the blade at about 24,000rpm to deliver 22.3ft^3 (0.631m^3) of oxygen at 5.0lb/min (2.27kg/min) and 70°F (21.1°C) with a pressure rise of 0.5in (1.13cm) of water. Under normal operating conditions the fan used 17W of power. A screen at the fan inlet prevented contaminants flowing through and the outlet was clamped to the cabin heat exchanger.

The design and assembly of the cabin heat exchanger was similar to the suit-circuit heat exchanger and was capable of rejecting about 500Btu/hr (527kj/hr) at a cabin temperature of 100°F (37.8°C). Under these conditions the cabin heat exchanger required a water flow of 0.5lb/hr (0.23kg/hr), which was manually controlled with adjustments to a needle valve.

The cabin pressure relief valve worked to relieve cabin pressure from sea level to the vacuum of space and to limit the upper value of pressure while the spacecraft was in orbit. It operated as a differential type at 5.5lb/in^{2}g (37.9kPag), +0.4 to -0.1lb/in^{2}g (+2.7 to -0.69kPa). The valve started to operate by repressurising the cabin during the descent when the ambient air pressure exceeded the cabin pressure by about 1lb/in^2 (0.69kPa). It incorporated a redundant path for both the control element and the relief valve and

LEFT From left to right, astronauts Shepard, Glenn, Schirra, Carpenter, Cooper, Slayton and Grissom, with Robert 'Bob' Gilruth of Space Task Group, examine the form-fitting contour couch which was itself an integral part of the life support functions of the Mercury spacecraft. *(NASA)*

a manual override allowed the astronaut to manually depressurise the cabin in the event of a fire or a build-up of toxic gases. A separate manual control allowed the valve to be locked closed after splashdown to prevent water from entering the spacecraft.

During the descent of the spacecraft the ground ventilation system was activated at an altitude of 17,000ft (5,180m), at which height both inflow and outflow valves were opened barometrically which allowed air to be drawn into the suit circuit by the suit compressor and to be vented to the cabin via the relief port in the suit pressure regulator. The outflow valve relieved this flow to the ambient environment to prevent pressure building up inside the cabin. The inflow valve was equipped with a snorkel that prevented water from being sucked into the suit circuit if the spacecraft submerged after splashdown. It incorporated three electrical switches which, when switched off, turned off the cabin fan, placed the suit circuit in emergency mode (by opening the emergency oxygen rate valve and closing the system shut-off valve), and de-energised the suit-fan circuit cut-off relay so that the suit compressor continued to function.

After the spacecraft landed it was possible for water – trapped in the recovery compartment where the inlet snorkel was located – to cause the snorkel ball to seat, thus preventing air from being drawn into the suit circuit. To prevent this, a negative-pressure relief valve was placed in the inflow line between the inflow valve and the suit compressor. The valve was operated in such a way that it relieved the vacuum in the line when the differential pressure reached 15in (38cm) below cabin pressure. At this point the valve would open allowing cabin pressure to relieve the line, releasing the ball float. The valve was designed to re-seat when the vacuum decreased to 2in (5cm) of water.

Both inflow and outflow valves were of the spoon type and spring-loaded in the open position, held closed by an aneroid assembly. During ascent to orbit the aneroid would expand and arm the valve with a detent pin and on descent the aneroid would retract and release the spring, closing the valve. The astronaut could override the system and manually open both valves.

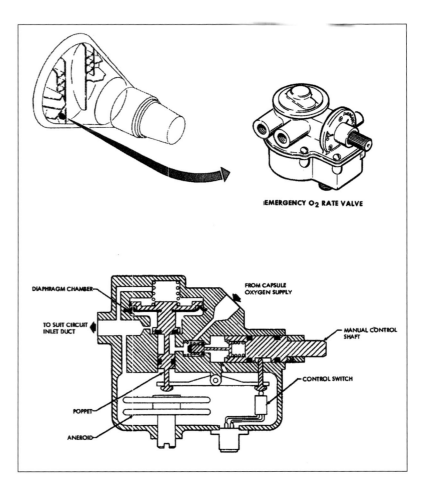

IEMERGENCY O₂ RATE VALVE

DIAPHRAGM CHAMBER
TO SUIT CIRCUIT INLET DUCT
FROM CAPSULE OXYGEN SUPPLY
MANUAL CONTROL SHAFT
CONTROL SWITCH
POPPET
ANEROID

Challenges

During the development phase several changes were deemed necessary. Originally the oxygen-supply tanks had a quick-disconnect, spring-loaded poppet valve, but those which tested clear were found to leak when the fully assembled system was checked out. The leakage was attributable to contamination in the metal-to-metal seals within the valve housing. A new valve design was finally chosen that had a tapered-shaft seal. Requiring a wrench to connect and disconnect the service line and to open and close the filler port, this functioned perfectly well throughout the programme.

Several other subsystem changes were necessary prior to flight operations, including deletion of a pressure switch that would have informed the astronaut when the primary oxygen supply was about to run out and that the secondary system was about to cut in. In numerous tests small quantities of aluminium housing were vaporised when the O-ring burned through and it was felt better

ABOVE Installed below and to the right of the astronaut couch, the emergency oxygen rate valve operated automatically when the suit pressure fell below 4lb/in² or when manually activated by the astronaut via an 'EMER O₂' handle. *(McDonnell)*

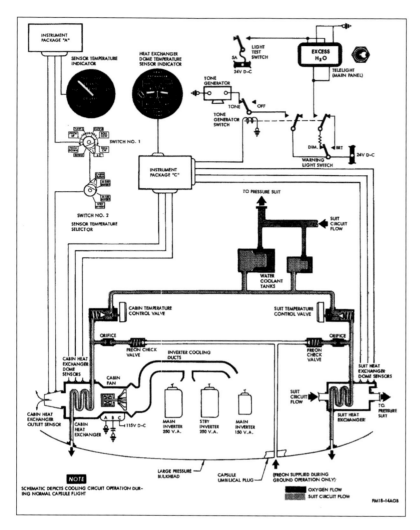

Schematic depicts cooling circuit operation during normal capsule flight

FM18-14AG8

to eliminate this altogether. The function was replaced with oxygen flow sensors. These used a thermistor downstream of each oxygen supply check valve in a configuration that would illuminate the 'O₂ QUANTITY' warning light with a switch on the control panel to show the status of the primary or secondary system.

Another change was introduced when the initial ECS design had a launch purge system that employed a tank of gaseous oxygen and a barometrically actuated purge valve to pump up the oxygen quantity and remove the mixed gas, oxygen/nitrogen atmosphere of the launch site. This would automatically initiate flow from the 1lb (0.45kg) supply when cabin pressure had decreased to an altitude of 10,000ft (3,048m) and the supply would be depleted when the altitude had decreased to an equivalent of 27,000ft (8,200m); but calculations showed this effective in only 65% of cases, and although it was used in the

Mercury Redstone flights it was removed for the orbital missions.

Just by removing this system engineers saved 4.75lb (2.15kg) of spacecraft weight which they replaced with an external purge fitting in the spacecraft hatch, activated after the astronaut was sealed inside. About 15lb (6.8kg) of oxygen was used to achieve a minimum purity of 98% oxygen prior to launch, removed from the cabin by a manual decompression port in the cabin pressure relief valve. Changes were also made to the suit-circuit purge technique to prevent nitrogen within the hoses leaking into the suit proper.

During the Little Joe-5A flight the cabin pressure decayed to zero and post-flight examination revealed that a piece of safety wire had lodged on the seat of the cabin pressure relief valve. Modifications were made by applying debris-protection screens over the cabin relief ports of the valve and an additional screen was added to cover the ambient or recovery compartment port of the valve. Spacecraft 13, 18 and 20 had a manual locking device that allowed the astronaut to positively lock the valve against water entering after splashdown.

Changes were also made to the water separator, a dry-sponge design, which had a tendency to become stiff and brittle, being changed for one of polyurethane that was flexible when dry. In addition a magnet was incorporated into the actuating piston. But in order to provide the astronaut with a check on separator actuation a switch was added to the instrument panel to allow manual operation, and two magnetic switches were hooked up to separate lights on the panel to indicate partial and full travel of the piston. Changes were also made to the suit pressure regulator, with several subtle changes and design alterations as well as inevitable testing with various materials, most of which were catered for by adding protective screens over the relief port openings in the valve case.

With the exceptions of spacecraft 16 and 20, a coolant quantity indicating system (CQIS) was fitted to provide information on the amount of coolant remaining by measuring the pressure decay of the small oxygen cylinder, as the absolute-pressure control valve maintained a constant pressure on the gas side of the positive expulsion bladder. This consisted of a 22in³ (360cm³) supply cylinder, a 500lb/in³

(3,226cm³) coolant-quantity transducer and a coolant-oxygen-pressure reduction control valve. The transducer indicated the percentage of coolant quantity remaining in the 39lb (17.7kg) capacity tank.

Early in the first manned development tests conducted in November 1959 the emergency oxygen flow-rate valve was prone to produce an insufficient flow for proper cooling of the astronaut. The microswitch which triggered emergency mode flow when pressure dropped to 4.0lb/in²g (27.6kPag) stuck repeatedly in the closed position until a spring was added to hold it open until actuated. Further troubles were encountered with the linkages connecting the emergency rate valve, the system shut-off valve and the rate handle in spacecraft 5, 7 and 8, difficulties accruing from the complicated rigging process.

Employing a Teleflex cable in compression to return the system from emergency to normal mode, the spring required a force of 40lb (18.1kg) to actuate. This had to be attached to a bellcrank on the rate valve and thence to the system shut-off valve. During the unmanned MA-4 (spacecraft 8A) test flight the rate valve partially opened to allow a high oxygen use rate, but to prevent this occurring without indication on a manned flight a microswitch would actuate between 20° and 25° of travel with flow initiated thereafter, the 20° of deadband compensating for the difficulties with valve rigging. A solid linkage replaced the Teleflex cable on spacecraft 16, 18 and 20.

Significant challenges faced engineers working the carbon dioxide and odour absorber, initially causing LiOH dusting problems at the absorber exit during vibration tests. The synthetic felt mat did not completely filter out the lithium hydroxide dust that could prove irritating to the throat and nasal passages, and a polyurethane foam was added. The LiOH quantity was increased from 4.6lb (2.08kg) to 5.4lb (2.45kg) to allow an absorber life extension from 34.5hr to 43.6hr. The activated charcoal was decreased from 1lb (0.45kg) to 0.2lb (0.09kg). Post-flight analysis of the MA-9 flight showed that only 73% of LiOH had been used, although an increase in carbon dioxide levels only began after 32hr 15min of flight. Detailed analysis revealed that on this flight

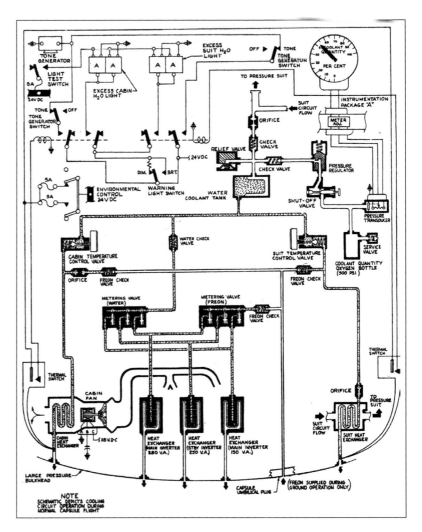

90% of the two upper bags of LiOH had been expended and 55% of the two lower bags.

Among other challenges, the outflow snorkel valve designed to prevent water entering the pressurised compartment after landing had to be redesigned, a lightly loaded diaphragm check valve replacing a screen-retained ball. To ensure a pure oxygen atmosphere in MA-6 a constant-bleed oxygen supply was introduced to the pressure suit circuit in excess of the metabolic requirements of the astronaut. This metered oxygen at a rate of 43.93in³/min (720cm³/min) to the pressure suit inlet duct, a system provided since the design capacity of the oxygen tanks was far greater than the duration of the mission. It was not installed for other missions.

O-ring seating in the ground ventilation inlet valve caused severe problems and on the unmanned MR-2 flight the cabin pressure decayed to zero when vibrations caused it to

ABOVE This schematic displays the operational layout and electrical switching for the cooling circuit in greater detail. Note the coolant quantity indicator on the instrument display and interconnectivity with the pressure suit circuit. *(McDonnell)*

open 57 seconds after lift-off on this ballistic flight. The problem was eventually solved by installing a new latching mechanism. Changes to the duct-mounted oxygen check valves located at the discharge side of both suit compressors were necessary when it was discovered that orientation to gravity – vital for their effective operation – was compromised until a positioning lug was installed. This was frequently found to be installed incorrectly. Moreover, light coil springs had to be added to the flapper discs to allow sufficient time for the check valves to change position.

Specific modifications were necessary too for the negative-pressure relief valve to prevent sticking, due to contamination of the O-ring seal lubricant, to the comfort control valve that manually controlled the rate of water supplied to the heat exchangers; to the high-pressure oxygen shut-off valve; and to the high-pressure

reducer in the secondary oxygen supply, which exhibited an unexplained decay during MA-6. Subsequent tests conducted over several weeks after the flight showed erratic results and the problem was only solved when slivers of O-ring material were found impacted against the sintered filter; hand-carried back to the subcontractor plant, no replication of the decay was observed after the O-ring was replaced.

By far the most difficult problem facing ECS engineers was the Freon check valve, corrosion with the initial valve configuration causing a change of materials from aluminium to stainless steel. The rejection rate was very high with the new valves because of contamination and sticking due to rough surfaces and poor finish. The specification was altered to stimulate a smoothness surface finish of at least 16. The smoothness factor reflects the amplitude of the number of waveform undulations in micro-inches. A valve in the cold-plate inverter cooling system failed in the closed position during the MA-6 mission and NASA changed the subcontractor, with a specification amended to 32 finish. This level of perfection cannot be obtained by grinding and has to be polished.

Test and qualification

The ECS test programme involved two separate systems: one for a complete integrated test and the second for subsystem and component testing. The integrated test involved 18 simulated 28-hour missions, the first ten under normal operation and the remainder with a range of emergency modes simulating a variety of different failures. These were conducted in a full-scale mock-up of a Mercury spacecraft. The simulator provided metabolic uptake and outgassing of carbon dioxide similar to an average astronaut with samples taken every 30 minutes. Key subsystems to monitor during this simulated re-entry phase were the cabin pressure relief valve, snorkel valve and demand regulator.

The total number of integrated system tests involved 488 hours under normal conditions, 16 hours under simulated emergency modes, and 29 hours simulating post-landing activity. One primary oxygen supply subsystem was tested under 36 mission cycles, of which 20 were

BELOW Subtle modifications for spacecraft 18 with changes to the Freon check valve and metering elements. *(McDonnell)*

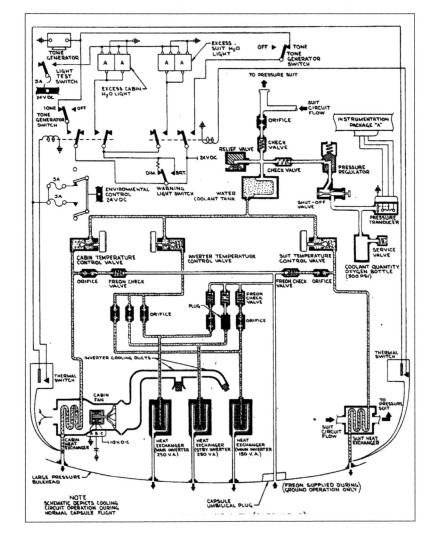

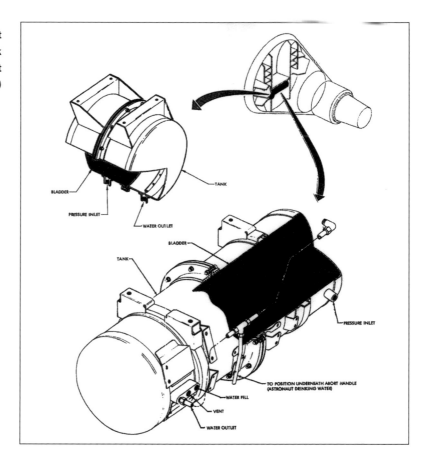

normal and 16 emergency simulations. Another was tested for 10 and 8 cycles respectively.

The qualification test programme involved a full simulation of vibration, acceleration, temperatures and acoustic profiles the ECS would experience in flight. Temperatures ranged from 200°F (93.3°C) to -20°F (-28.9°C), with 50 hours' exposure to sand and dust. Mechanical shock simulation exposed the full ECS configuration to 15g for 11ms in all axes and directions, and a linear impulse of 100g for 11ms. Five complete sets of equipment were used for the ECS qualification tests, which included burst-proofing of tanks and 100g shock tests. Two of the sets were used for systems tests and three for component-level qualification.

Critical to mission success was the integrity of the oxygen tanks, eight of which were oxygen-aged at 8,500lb/in²g (58,607kPag) for 150 hours with each tank subject to 10,000 hydrostatic pressure cycles or to failure. Each tank was then subjected to a complete pressure cycle and taken to burst pressure of 16,700lb/in²g (115,146kPag). Seven oxygen tanks were then oxygen-aged for 1,000 hours and then pressurised to burst level. Fifteen tanks were also given 1,000 hours' oxygen-ageing and taken through 10,000 pressure cycles before being subject to burst test.

Beginning in November 1959 manned tests were conducted at the AiResearch plant. Lasting a week they involved a McDonnell test pilot for altitude simulation so as to evaluate the full operability of the ECS with a man in the loop. A physician from the Lovelace Clinic was on hand to monitor the tests and to assess the physiological condition of the test subject. Various cycles revealed ways in which

RIGHT The water separator collected condensed water from the heat exchanger and consisted of a sponge within a rigid container squeezed by a pneumatic piston. *(McDonnell)*

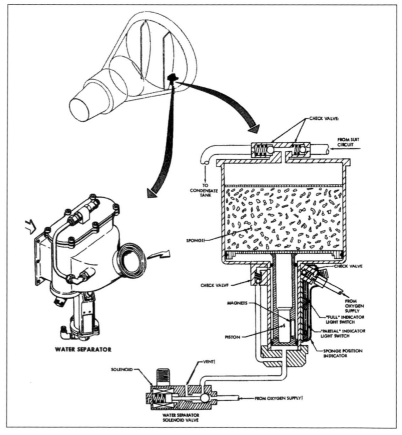

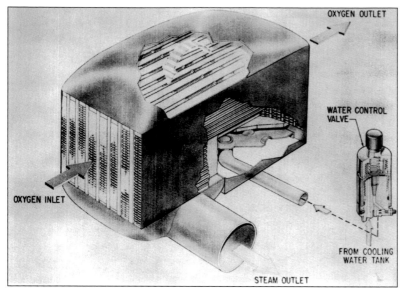

OXYGEN OUTLET

WATER CONTROL VALVE

OXYGEN INLET

FROM COOLING WATER TANK

STEAM OUTLET

ABOVE A schematic showing the evaporator for the zero gravity and water control valve. *(McDonnell)*

BELOW This time-chart of the temperature history for the 22-orbit MA-9 flight shows erratic values early and at sporadic periods in the mission. The black bars indicate when the spacecraft was in darkness. *(NASA)*

settings and the precise configuration could be improved and potential failures averted. The total cumulative test duration was 24 hours and it paved the way for a second set of manned tests on 18 March 1960, with special emphasis on integration of suit and ECS testing. The modifications and improvements were incorporated into all capsules starting with spacecraft 5.

McDonnell conducted their own separate set of tests consisting of 12 runs in a wide range of operating modes, focusing on biomedical monitoring and on simulated telemetry using actual spacecraft hardware. A significant problem was found with the relief diaphragm of the suit pressure regulator but this was corrected prior to the completion of the McDonnell test programme in October 1960.

Thereafter the equipment was shipped to the Air Crew Equipment Laboratory at the Naval Air Material Center and installed in an altitude chamber, where it was used by NASA to train the astronauts on its functionality and operation.

Flight hardware was subject to detailed testing at McDonnell's St Louis facility and when it was delivered to NASA at Cape Canaveral it was subject to sea-level and altitude tests. During launch preparation the spacecraft was installed in the altitude chamber at Hangar S and underwent a further run of verifying trials and tests. The altitude test was the last one conducted prior to mating of the spacecraft to the launch vehicle on the pad, although some checks were made after integration as a reassurance that none of the other preflight preparations had affected the performance of the ECS.

One test considered vital was a pressure integrity check where the hatch was installed and the spacecraft pressurised to $5.0lb/in^2g$ (34.47kPag) (19.7lb/in²/135.8kPa) with a rotameter supplying make-up oxygen for a monitoring period of one hour. The leakage rate observed was converted to the rate of pressure decay from $3.0lb/in^2g$ (26.7kPag) for the free gas cabin volume of $52ft^3$ (1.47m³). To verify this leak rate on the day of flight, the cabin was pressurised to $3.0lb/in^2g$ and the decay rate measured for between four and eight minutes. The actual values read on launch day were compared with graphs prepared from the previous sampling period prior to launch day.

Flight results

The only noteworthy event with Little Joe flight tests was the 5A launch with spacecraft 14 where the cabin pressure decayed to zero due to a wire lodged on the seat of the cabin pressure relief valve. The ground ventilation inlet valve on spacecraft 5 launched by MR-2 vibrated open and the cabin depressurised via the negative-pressure relief valve, which caused the suit circuit to command the emergency rate. A snubber intended to absorb vibration loads had been inadvertently removed before flight. Water also entered the capsule after splashdown via the cabin outflow snorkel, which was modified. A

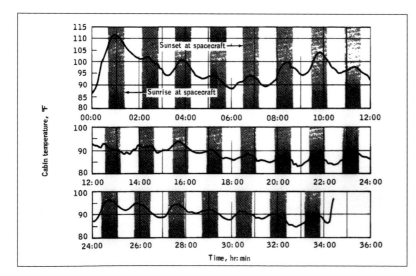

significant quantity of Freon-114 was absorbed by the charcoal bed of spacecraft 7 launched on MR-3, but this was later found to have been caused by the preflight manned altitude chamber tests.

Of the Mercury Atlas flights, spacecraft 8A launched on MA-4 experienced an abnormally high oxygen-use rate, caused by a partially open emergency-rate valve that had been loosened by vibration during ascent. Launched with a primate on board, spacecraft 9 (MA-5) experienced a sharp rise in temperature due to preset flow control valves not being at a sufficiently high rate, which brought about a decision to return it to Earth after two orbits. A man would have adjusted it correctly. Uniquely, the MA-6 mission carried spacecraft 13 with a suit circuit constant-bleed orifice rather than a demand regulator as on all other missions. However, the check valve between the Freon-114 and water inlets to the inverter cold-plates failed, preventing water from entering the cold-plates and leading to higher inverter temperatures.

On spacecraft 18 (MA-7) the coolant-quantity indicating system failed, but the CQIS reading was obtained by telemetry and the sensor was removed for the last two flights. Suit cooling was a big problem, as it had been on spacecraft 13, but the results of tests were not in by the time MA-7 flew. It was later determined that the heat exchanger duct temperature was not a good place to get a control parameter and that the steam temperature in the heat exchanger interpass or dome area was the better location. Consequently the control parameter reading was relocated for spacecraft 16 (MA-8). But even on this flight the astronaut had to continuously increase the coolant flow setting and only got a satisfactory temperature two hours after launch, a problem later resolved by changing the lubricant.

Arguably the biggest challenge to the reliability and functionality of the ECS came with the 34-hour mission of spacecraft 20 (MA-9). As recorded earlier, several changes were made to this spacecraft in advance of its flight and as a result of development and improvement. To evaluate the cabin cooling circuit the astronaut turned off the cabin coolant water

and fan at 6hr 22min elapsed time. Efficiency was to be measured by a temperature probe located at the cabin heat exchanger outlet and this showed a temperature increase to within 10°–15°F (5.55°–8.33°C) below the indicated cabin temperature, which maintained a consistent 90°–95°F (32°–35°C). As a result cabin cooling was not considered necessary during the extended power-down phase where the spacecraft was in drifting flight.

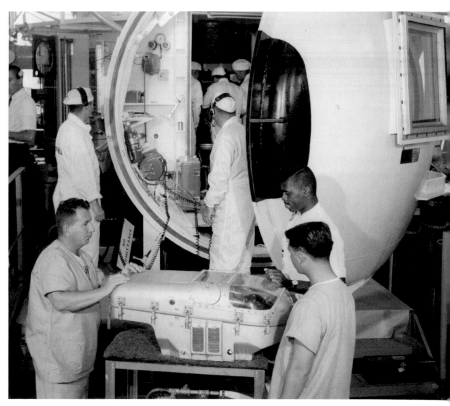

Attitude control systems

The precise orientation of the spacecraft was crucial to the safe return of a capsule and its occupant, because the fixed-axis retro-rockets had to be correctly aligned for the burn to bring the spacecraft out of orbit with the correct velocity vector. The three orthogonal axes (x, y and z) set the reference frame of the spacecraft, aligned with the pilot facing forward, shoulders straight and level, the blunt forebody being behind, the recovery section being in front. Readers will note that this was contrary to the definition of the structural elements of the spacecraft, where the forebody was the aft-facing heat shield. The attitude reference frame assumed the normal position of a pilot, head up and facing forward.

The z-axis was aligned with the longitudinal centreline of the spacecraft (around which clockwise or anticlockwise roll was measured), the x-axis was at right angles (determining pitch up or down) and the y-axis was at right angles to the pitch axis (determining yaw left or right). Uniquely for this spacecraft, instead of 'plus' or 'minus' for extremes of the x and y axes, 'up' and 'down' or 'left' and 'right' were used for relevant placements as viewed by the pilot seated and looking toward the recovery compartment.

Attitude stabilisation was accomplished with the Automatic Stabilization Control System (ASCS) in conjunction with the horizon scanners and the Reaction Control System. Bell Aerosystems was responsible for the thrusters,

the tiny rocket motors that would maintain attitude. The company had been involved in thrusters for attitude control since 1953, when it provided the wingtip rockets for the Bell X-1A rocket-powered research aircraft, and had also provided attitude thrusters for the hypersonic X-15. The concept of using thrusters for this purpose had been pioneered by the X-series research aircraft, and NACA had done much work on their development and use with feedback input and side-stick controllers.

But the philosophy of redundancy and the duplication of a parallel set of subsystems to allow safe recovery of the spacecraft and its pilot were crucial when it came to attitude control. A redundant, or backup, system to the ASCS was provided in the form of the Rate Stabilization Control System (RSCS), allowing the astronaut to use a 'rate stick' in the event the primary system failed. As further backup there were means for the astronaut to conduct a visual check of roll, pitch and yaw.

ASCS

The ASCS was designed to take control of the launch vehicle from the time of separation from the vehicle to the deployment of the landing parachute. It was to provide output signals for display, recording on-board and telemetering of three-axis information, a discrete signal at 0.05g longitudinal acceleration during re-entry, and an attitude signal sensor for use in the retrograde firing interlock circuit.

Associated equipment for the ASCS included rate gyros, horizon scanners, reaction controls, communication system telemetry, devices for the display of spacecraft attitude and other devices for generating capsule signals for certain mission events. But the control system was also designed to map operations in such a way as to minimise the use of the limited RCS propellant in limited-cycle oscillations.

Four modes of operation were designed into the ASCS: damper mode; orientation mode; attitude-hold mode; and re-entry mode. In the electro-mechanical world of the late 1950s, these were the equivalent of separate computer programmes for specific phases of the mission or a particular mode of operation. In this, Mercury was a precursor to the computerised

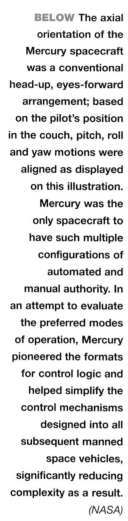

BELOW The axial orientation of the Mercury spacecraft was a conventional head-up, eyes-forward arrangement; based on the pilot's position in the couch, pitch, roll and yaw motions were aligned as displayed on this illustration. Mercury was the only spacecraft to have such multiple configurations of automated and manual authority. In an attempt to evaluate the preferred modes of operation, Mercury pioneered the formats for control logic and helped simplify the control mechanisms designed into all subsequent manned space vehicles, significantly reducing complexity as a result.
(NASA)

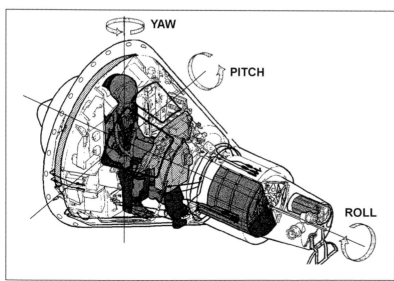

YAW

PITCH

ROLL

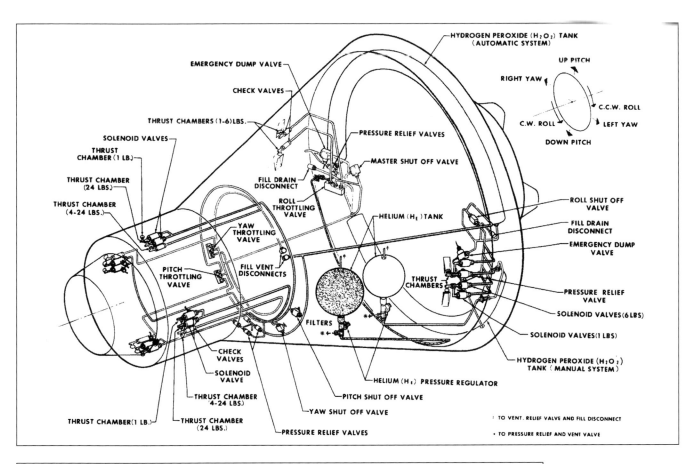

HYDROGEN PEROXIDE (H_2O_2) TANK
(AUTOMATIC SYSTEM)

UP PITCH
RIGHT YAW
C.C.W. ROLL
C.W. ROLL
LEFT YAW
DOWN PITCH

EMERGENCY DUMP VALVE
CHECK VALVES
THRUST CHAMBERS (1-6)LBS.
SOLENOID VALVES
THRUST CHAMBER (1 LB.)
THRUST CHAMBER (24 LBS.)
THRUST CHAMBER (4-24 LBS.)
PRESSURE RELIEF VALVES
MASTER SHUT OFF VALVE
FILL DRAIN DISCONNECT
ROLL THROTTLING VALVE
HELIUM (H_e) TANK
YAW THROTTLING VALVE
PITCH THROTTLING VALVE
FILL VENT DISCONNECTS
ROLL SHUT OFF VALVE
FILL DRAIN DISCONNECT
EMERGENCY DUMP VALVE
THRUST CHAMBERS
PRESSURE RELIEF VALVE
SOLENOID VALVES(6LBS)
SOLENOID VALVES(1 LBS)
HYDROGEN PEROXIDE (H_2O_2) TANK (MANUAL SYSTEM)
FILTERS
CHECK VALVES
SOLENOID VALVE
THRUST CHAMBER 4-24 LBS.)
PITCH SHUT OFF VALVE
HELIUM (H_e) PRESSURE REGULATOR
YAW SHUT OFF VALVE
THRUST CHAMBER(1 LB.)
THRUST CHAMBER (24 LBS.)
PRESSURE RELIEF VALVES

† TO VENT. RELIEF VALVE AND FILL DISCONNECT
* TO PRESSURE RELIEF AND VENT VALVE

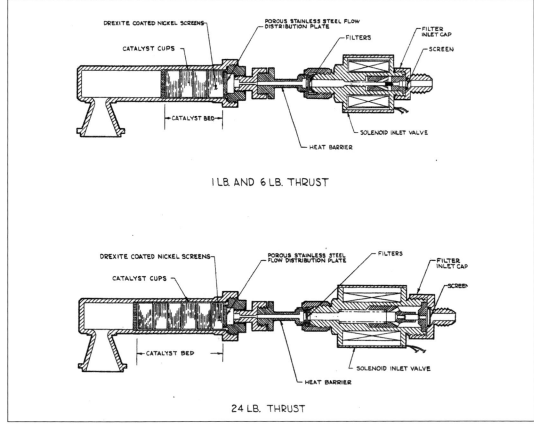

DREXITE COATED NICKEL SCREENS
POROUS STAINLESS STEEL FLOW DISTRIBUTION PLATE
FILTERS
FILTER INLET CAP
CATALYST CUPS
SCREEN
CATALYST BED
SOLENOID INLET VALVE
HEAT BARRIER

1 LB. AND 6 LB. THRUST

DREXITE COATED NICKEL SCREENS
POROUS STAINLESS STEEL FLOW DISTRIBUTION PLATE
FILTERS
FILTER INLET CAP
CATALYST CUPS
SCREEN
CATALYST BED
SOLENOID INLET VALVE
HEAT BARRIER

24 LB. THRUST

ABOVE Relevant to the axial orientations of roll, pitch and yaw (top right), the reaction control system provided a set of individual thrusters based on the decomposition of hydrogen peroxide over a catalyst to produce a reaction from the combustion process and effect a rotation of the spacecraft as a result. *(McDonnell)*

LEFT The two separate thruster configurations provided three levels of fixed thrust identified here with variable size and catalytic bed length as shown. *(McDonnell)*

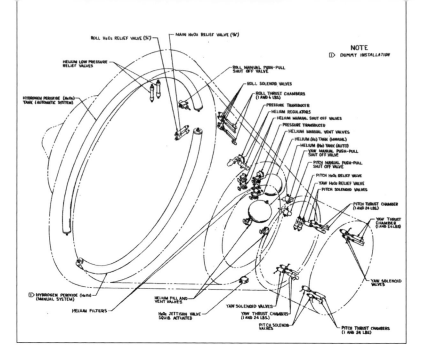

programmes utilised in a very limited degree by Gemini and to a greater extent by Apollo.

In addition to these modes the ASCS was designed so that the astronaut could select between manual fly-by-wire (FBW) and auxiliary damping modes. In the FBW mode the astronaut had authority over the automatic RCS thrusters as the rate damping mode was switched on and control functions were energised so that he could use a side-stick controller linked to stick position potentiometers. The auxiliary damping mode provided only rate damping and the automatic and FBW functions were disengaged. There was no provision for the astronaut to intervene by manually stepping the ASCS automatic sequencing.

While some variability was built into the ASCS, the general sequence of operations was that required for a standardised mission but also included rate-damping in the event of an abort and rate-damping and orientation to the desired attitude during high-altitude aborts or in normal missions. Otherwise the ASCS controlled the orientation of the spacecraft throughout orbital flight with respect to the local vertical (astronaut's head up) and orientation of the spacecraft to specified pitch angles (-24° to -60° blunt-end up) prior to retrofire. It held attitude throughout the retrofire burn of the solid propellant rockets and controlled reorientation to selected re-entry attitude after the firing. Beginning with the build-up of drag at the 0.05g point it was to provide a steady rate of roll of

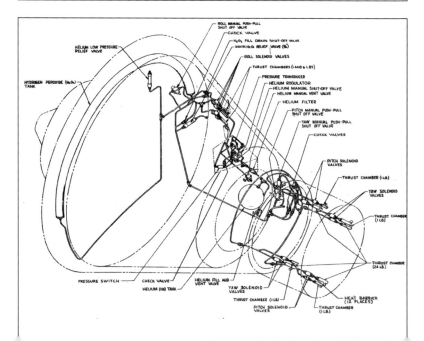

about 10–12°/sec until disengagement when parachute deployment was initiated by a signal from the barostat.

The rate gyros were installed to sense and measure the magnitude of spacecraft rotational rates and the 0.05g sensor was to sense accelerations for the initiation of re-entry sequences. The attitude reference system consisted of a vertical and a directional gyro, which sensed pitch, roll and yaw attitude excursions along with signal inputs from the horizon scanners. Pitch and roll outputs of the horizon scanners were utilised to precess the gyros such that their spin axes could be maintained in the normally erected position relative to the moving local vertical axis.

Before launch the vertical and directional gyros were torqued so as to erect their spin axes to any desired orientation relative to the ascent trajectory. During ascent, but prior to escape-tower jettison, the vertical gyro spin axis was erected to the horizon scanners. After tower jettison, and for a period of approximately five minutes to the point of separation from the launch vehicle, both the vertical and directional

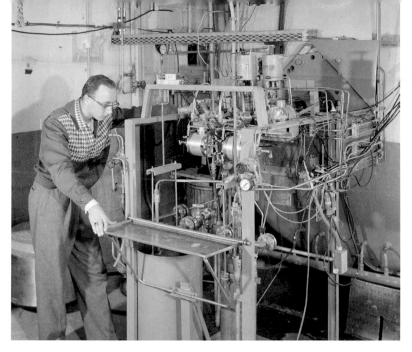

ABOVE Considerable research went into the development of a hydrogen peroxide reaction control system for the X-15 hypersonic research aircraft and the Mercury spacecraft. This test rig at the Lewis Research Center was employed during 1959 on development of the Mercury thruster system and its storable propellant. *(NASA)*

BELOW This schematic of the automatic RCS subsystems shows the slaved propellant delivery lines from the upper semi-toroidal hydrogen peroxide tanks to the various thrusters as indicated along the bottom. *(McDonnell)*

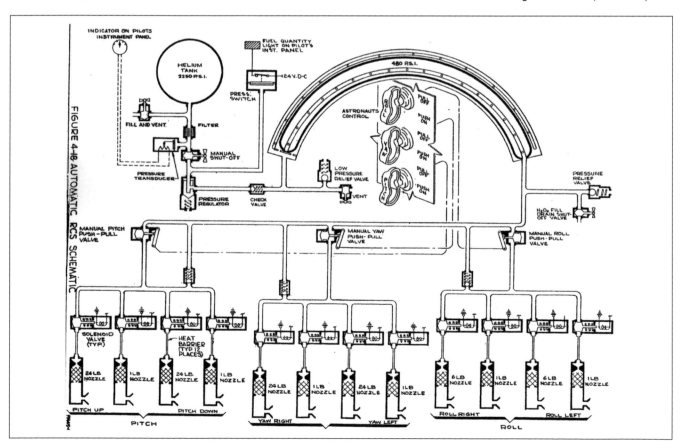

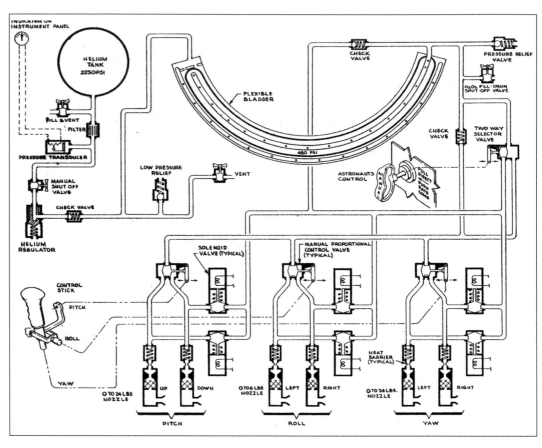

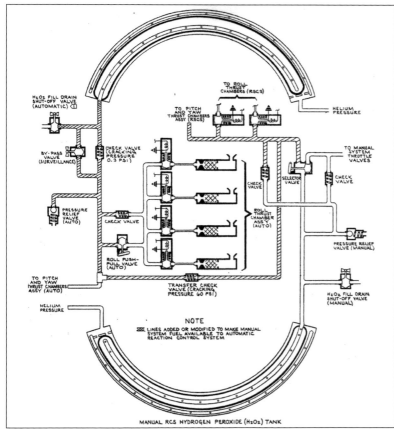

(roll gimbal only) gyros were slaved to the
horizon scanners. During normal orbital flight
both gyros were slaved to the horizon scanners
during their intermittent operation as generated
by signals from the data programmer.

For the ten minutes prior to retrofire both
gyros were continuously slaved to the horizon
scanners. At retrofire the gyros were not slaved
to the horizon scanners, which were taken
off line and inoperative. At retro-pack jettison,
and until the 0.05g sensor responded to this
acceleration (retardation), the gyros were
once again slaved to the horizon scanners.
At 0.05g the gyros and the horizon scanners
were taken off line and the spacecraft pitch
and yaw angular rates would be reduced to
a value of 0.8°/sec or less, while a steady
state roll of 10–12°/sec was maintained until
disengagement of the ASCS at main landing
parachute deployment at an altitude of 10,000ft
(3,048m). The ASCS had a total weight of 59lb
(26.7kg) and was designed to provide an almost
hands-free operation.

A brief description of its use during a typical
mission emphasises the pre-programmed

nature of the typical operating plan for the way the spacecraft was planned to run during orbital flight. Three switches were provided in conjunction with operation of the ASCS by the pilot: the 'GYRO' switch, the 'AUTO/RATE COMD' switch and the 'NORM-AUX DAMPO-FBW' switches located on the astronaut's left-side instrument board. With the 'NORM-AUX DAMPO-FBW' in the 'NORMAL' position and the 'AUTO-RATE COMD' in 'AUTO', stabilisation operated in a completely autonomous manner that required no further assistance from the astronaut. In the 'FLY-BY-WIRE' position the automatic feature was disabled and 24V DC power applied to the limit switches on the astronaut's control stick.

In this configuration, stabilisation was accomplished through an electro-mechanical arrangement by motion of the control stick in the desired plane. Low and high thrust actuation occurred at approximately 30% and 75% of full travel, 3.9° and 9.8° for roll, pitch and yaw. The 'AUX DAMP' position disabled both automatic and fly-by-wire functions, permitting rate damping as a single feature. The 'GYRO' switch had three optional positions – 'CAGE', 'FREE' and 'NORMAL'. In the 'CAGE' position the attitude gyros (see below) were mechanically caged and the horizon scanner (see below) slaving function was disabled. In the 'FREE' position the attitude gyros were uncaged but the horizon scanner slaving functions remained disabled. In the 'NORMAL' position attitude gyros were uncaged and slaving was allowed for the horizon scanners. The 'AUTO/RATE COMD' switch provided a method of energising either the RSCS or the ASCS systems as desired. In the 'RATE COMD' position the attitude gyros and slaving circuits remained energised, although they were not used to control the attitude of the spacecraft.

In normal operation during a trouble-free flight the entire sequence of ASCS operation was divided into eight phases. Phase A required gyro-slaving to the horizon scanner pitch and roll outputs during ascent so as to minimise the possibility of gyro errors, which could build up during the boosted phase to orbit. Phase B began after the capsule separated and when a five-second signal commanded the ASCS to provide rate damping, to null out any oscillations

imposed by the action of separating from the launch vehicle adapter and to inhibit tumbling.

Phase C began five seconds after the start of rate damping and placed the capsule in its appropriate attitude by turning it around and placing the pilot's back to the direction of flight. This was achieved in a 180° anticlockwise yaw

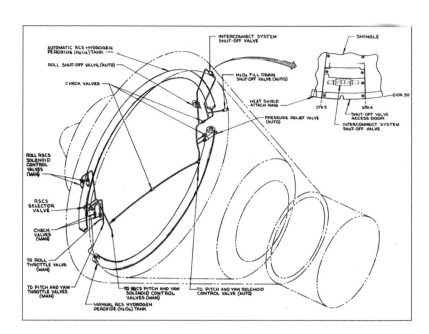

ABOVE The interconnection of automatic and manual subsystems as laid out in spacecraft 9, which flew the MA-5 mission. (McDonnell)

BELOW The automatic stabilisation control system (ASCS) and the rate stabilisation control system (RSCS) used a common set of support equipment such as horizon sensors and potentiometers as depicted in this cutaway, which identifies key elements and where they were situated within the spacecraft. (McDonnell)

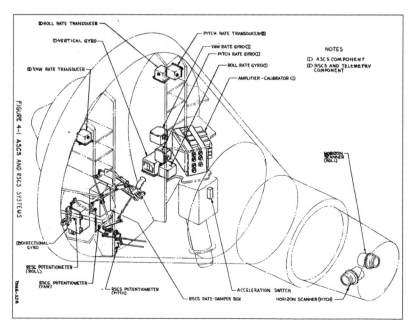

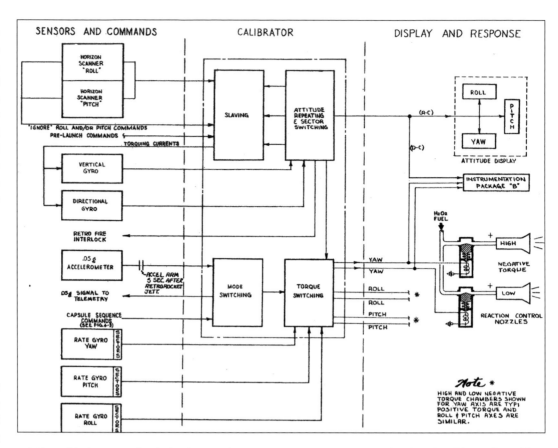

rotation with the capsule pitched nose-down to a -34° angle. This was achieved within 30 seconds and placed the spacecraft in the correct orientation for retrofire, should that be needed for an emergency return to Earth. Pitch, roll and yaw gyro-slaving to the horizon scanners was provided during Phase C and for the first 4.5 minutes of Phase D.

Phase D covered the period while the spacecraft was in orbit and maintained the pitch-down attitude, an orientation ready for immediate in-orbit abort. In this phase the attitude gyros were slaved to the horizon scanner as long as the ASCS 'GYRO' switch remained in the normal position. During the orbit phase manual control and fly-by-wire could be selected as desired. Rate damping became an option under manual control conditions by the astronaut positioning the ASCS 'MODE SELECT' switches and RCS controls. Through switch manipulation, rate damping was provided by either the ASCS or the RSCS. Rate gyro run-up was continued throughout this phase. If the capsule drifted from orbit attitude beyond the limits of the retro-interlock sector switches, automatic return to orientation mode would occur at ±12° in pitch and ±30° in yaw and roll.

Phase E covered the orbital period but assured rate gyro run-up, which was automatically switched on ten minutes before retrofire. The astronaut could change any one or

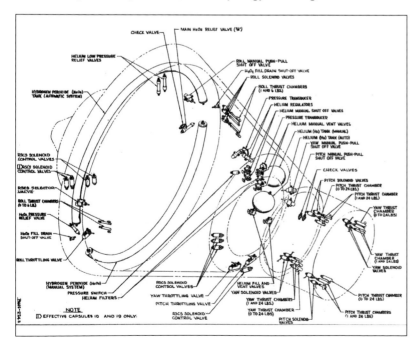

all three attitude orientations maintained by the ASCS by changing the space reference plane, or planes, of the attitude gyros. To maintain a new or reset reference plane the horizon scanner slaving command was stopped by placing the gyro switch in the 'FREE' position. However, new reference planes had to be established by the astronaut while the ASCS was engaged by placing the gyro switch in the 'FREE' position, manually turning off the ASCS fuel in the axis or axes affected, using manual control to position the capsule before caging and then uncaging the gyros. The ASCS could then be returned to fully automatic operation in any axis with the exception of the horizon scanner slaving. To utilise that the capsule attitudes were required to be within the observation range of the scanners and the gyro switch had to be placed in the 'NORMAL' position.

Phase F involved holding retrofire attitude during the retrograde burn with the horizon scanner slaving discontinued. The retro-rockets were fired 30 seconds after the attitude command, and during the burn period the ASCS utilised high torque action to hold the capsule within 1° of the required pointing angles in all three axes. The retrograde rocket firing command and the ASCS high torque switching command occurred simultaneously. The retrograde burn phase was completed in 20 seconds and the high torque switching command was held for 23 seconds. On retrograde package jettison the ASCS automatically pitched the capsule to the post-retrofire attitude, initiating Phase G in preparation for re-entry drag. The ASCS returned to the orientation mode with constant scanner operation to accurately maintain re-entry attitude.

Phase H was triggered by the onset of re-entry, sensed by the 0.05g accelerometer switch that turned off the attitude gyro power. During this period the ASCS maintained a constant roll rate of 10°–12°/sec to minimise the splashdown dispersions. ASCS operations continued under this phase until main parachute deployment, at which point all ASCS power was removed.

During all eight phases outlined above the astronaut could opt for manual, fly-by-wire or RSCS stick-steering control and then return

to normal ASCS 'MODE' command. The design philosophy here was that the astronaut can become an intelligent backup to the fully automated system. Engineered as a reliability augmentation concept, this led to provision for gyro caging and other optional switching features designed to broaden operating flexibility for the astronaut and for the spacecraft under automated or semi-automated control. Alluded to earlier, the four control modes (automatic, fly-by-wire, rate-stick and direct with damper control) ensured that this was attained during mission operations.

RSCS

The Rate Stabilization Control System operated completely independently of the ASCS and was designed for redundant rate damping, being programmed to respond like an aerodynamic reaction to flight in the atmosphere. It provided the astronaut with a hand-stick controller with which he could perform in control-stick-steering mode. This derived a lot from the development of such a system for the X-15, which was just about to

ABOVE Honeywell provided much of the equipment for control logic in the ASCS; here technician Gerry Borasch provides scale to the electronics box, which contains 3,000 separate connections. *(Honeywell)*

enter flight-test phase. The concept of dual-redundant flight control operation would set the ground rules by which all future spacecraft would be designed.

The RSCS allowed the astronaut to operate like the pilot of a conventional aircraft in that movement of the stick controller provided capsule angular rates approximately proportional to, and corresponding with, stick deflection. The RSCS included a rate damper, three control-stick position potentiometers, the addition of six solenoid control valves and a fuel selector valve to the manual Reaction Control System, the attitude-indicating system pitch, roll and yaw rate transducers, plus the necessary mode select switch, connectors and associated wiring.

The use of the RSCS in the event of a failure in the ASCS after retrofire required a constant roll capability. In that eventuality the pilot would initiate a constant roll rate of 7°/sec, effected

by closing the ASCS 0.05g acceleration switch. This provided stabilisation of the spacecraft during the re-entry phase without requiring manipulation of the hand stick controller by the astronaut, who would be experiencing g-forces.

The RSCS would be activated by the astronaut selecting the rate command position of the 'AUTO-RATE' command mode select switch and by depressing the manual fuel control handle. This allowed the H_2O_2 thruster fuel to flow from the fuel selector valve to the solenoid valves. When the 'rate command' mode was selected electrical power was applied to the rate damper electronics, which would receive rate error information from the rate transducers and rate command signals from the hand controller linkage. These inputs were summed by the rate damper summing preamplifier-demodulator, where angular rates were required to exceed a dead-band zone of ±2°/sec before an output signal could be transmitted to torque switching logic relays. On receiving a signal for corrective positive and negative commands, the appropriate logic relay would energise the positive or negative solenoid valve and command the appropriate thruster to fire.

Horizon scanners

The concept and technology for infrared horizon sensors was developed by Barnes Engineering Company of Stamford, Connecticut, from a general research programme for space vehicle instrumentation. This company provided the sensors for the Mercury spacecraft.

Stabilisation was provided by two horizon sensors continuously scanning to detect the thermal horizon between the nearly absolute zero temperatures of outer space and the warmth of the Earth's troposphere. This horizon is the best available stable reference indicator for establishing a vertical to the Earth below. Using this reference, the sensors generated electrical signals that were used to control the spacecraft's pitch and roll gyroscopes and thrusters. Each sensor contained a detector for converting received infrared radiation to an electrical signal, a rotating prism for scanning the detector across Earth and space, and a transistorised electronics system for processing

BELOW A product of the electro-mechanical age, infrared horizon sensors for the Mercury capsule were provided by the Barnes Engineering Company and installed by McDonnell in final assembly of the spacecraft. *(Barnes Engineering Co)*

the detector's electrical output signal and converting it to the form of correctional signals.

The Mercury horizon scanner incorporated two identical scanning units and its purpose was to provide a roll and pitch reference during the orbital phase of a normal mission. The scanners each produced an output signal that slaved the ASCS attitude gyroscopes to the proper angles upon command from an external programmer. All the major components of the scanner were mounted on a large circular plate and included the scanning prism assembly, the prism drive, infrared detector, electronics, synchronous switches, electrical connector and a cover. The circular plate was flange-mounted so that the scanning prism compartment projected into space outside the spacecraft.

The electronics were completely transistorised and the various functional sections were fabricated on separate printed circuit boards, three of which were enclosed in the shielded housing fastened to the four posts mounted on the circular plate. The four posts with attached boards could be replaced as a single unit, or individual boards could be replaced as required. The horizon scanner was highly compact, weighing a mere 3.02lb (1.37kg) and being 6.15in (15.6cm) in length and 5.28in (14.7cm) in diameter.

The scanner had a number of special features, including a centrifugally activated shutter which prevented solar radiation from dwelling on the detector resulting in damage during periods when the scanner was static. Also, a special circuit could be used to disconnect error signals from the RCS thrusters when the presence of the Sun in the scan path, or the loss of horizon, would have resulted in erroneous error signals. A big advantage was that a single power source providing 110V, 400-cycle, 3.2VA was required to operate the entire system, this highly regulated power supply in the system eliminating the need for bulky batteries usually required to bias the infrared detector.

Operation of the horizon scanners depended upon the infrared radiation received from the Earth compared to the essentially zero radiation from space, and these differences provided a sharp radiation discontinuity at the horizon. The scanner used this discontinuity for both night

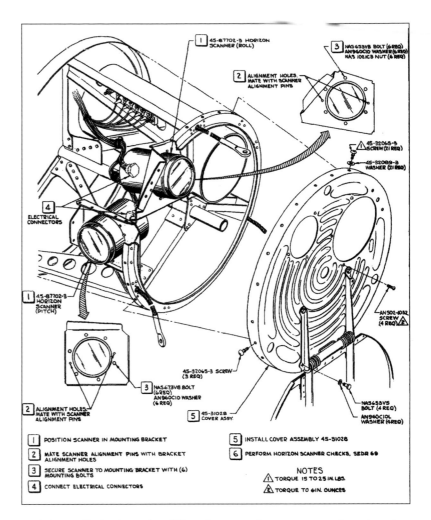

and day vertical referencing. When the capsule was orientated so that the Earth was present in its scanning path there were generally two points where the scan intersected the horizon. The scanner detected this change in radiation level and bisected the included angle from itself to the horizon points. It then compared the direction of the bisector with that of a fixed reference in the capsule and generated linear error signals proportional to the angle between the bisector and the fixed reference. These error signals in roll and pitch were used to slave the ASCS attitude gyroscopes.

The discontinuity observed by the scanner was approximately equal to that from black bodies at -459°F (-273°C), absolute zero, and -100°F (-73°C) respectively, and the radiance difference was approximately 0.00046W/in² (0.003 W/cm²) steradian. The location of this gradient was sharply defined and is much larger than any others encountered during the scan cycle. However, some sharp radiation gradients do exist and these

ABOVE The infrared horizon scanners were located in the forward section of the recovery compartment and just beneath the antenna fairing, as shown in this diagram providing instruction on installation. *(McDonnell)*

could be found at cloud edges, in topographical irregularities on the surface of the Earth and at the terminator line between night and day. These changes or irregularities could be filtered out so that the horizon gradient was the only one detected by the system.

Selective filtering was accomplished since most of the reflected solar radiation fell in the spectral range of 0.2–2.0 microns, while the radiation emitted by the Earth and the troposphere is at wavelengths longer than 5 microns. Filtering was accomplished by a germanium prism and a field lens in front of the detector; germanium sharply cuts out all radiation wavelengths shorter than 1.8 microns while transmitting uniformly radiation in the range 1.8–20 microns. By using this filter it was possible to remove more than 90% of reflected solar radiation, and signal clipping techniques in the electronics were able to effectively remove any residual effects.

The infrared detector was fixed to the centre of the circular plate and its field of view extended through the circular opening in the centre of the scanning assembly. The detector's field of view was 2° by 8° and the presence of the scanning prism had the effect of deflecting it 55° from the normal, therefore the apex angle of the scanning cone was 110°. During operation the drive system rotated the scanning prism and the detector field scanned the field of view through the conical pattern described earlier. Because varying amounts of radiation strike the detector during various portions of the scan cycle, the amplitude of the detector output changed accordingly. This output signal was processed by the electronics system and the error signal output produced was available at the electrical connector.

Strongly related to the prism drive system, the reference signal generator was the fixed reference frame against which the detector horizon signals were compared. The output of the generator was the square wave signal at a frequency of 30 cycles per second. The signal was triggered by the interaction between a magnetic pickup coil and a semi-circular steel vane embedded in a slot cut into the surface of the scanning prism assembly gear. A pickup was mounted so that the end of its magnetised core came close to the surface of the vane. As the scanning prism assembly turned, the ends of the vane passed the end of the magnetised pickup coil core, generating the reference pulse. The pulse was converted to a phase-locked 30-cycle square wave by an electronic network.

The Sun shutter consisted of a pair of spring-loaded metal slides that fitted into opposed transverse slots through the tube section of the scanning mirror assembly. When this assembly was not rotating spring tension pulled the two slides together and the detector field was obstructed. When the scanning mirror was turning the centrifugal force on the slides was sufficient to open the shutter.

The infrared detector was a thermistor bolometer with its active element immersed in the germanium lens. The active element here was a rectangular flake of thermister material that was connected in a bridge circuit with a similar compensating flake shielded from radiation. The two flakes were oppositely biased

BELOW The rate gyro installation diagram from the Mercury spacecraft servicing manual identifies the location of the boxes in the pressurised compartment. *(McDonnell)*

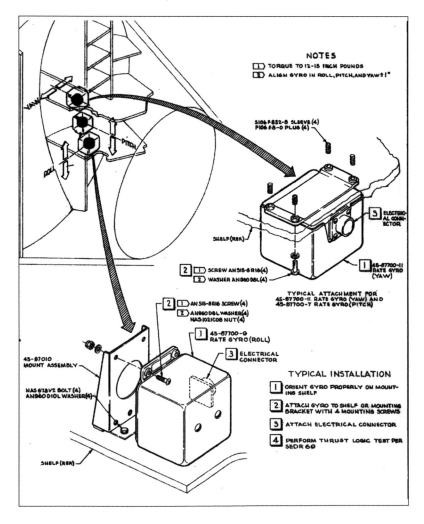

and their junction was connected to the input of the preamplifier. By immersing the active element in the rear surface of the germanium lens the overall level of detection could be increased by a factor of about 3.5 over an element that was not immersed. The material in the thermister flake had a high negative temperature coefficient of resistance, so when the temperature of the material increased the resistance of the flake decreased.

Because the surface of the thermister flake was blackened it absorbed impinging radiation and its resistance was decreased. When the shutter was closed both flakes in the detector bridge were at the same temperature, but because both had the same linear characteristics their resistance was the same. Gradual variations in ambient temperature changed the resistance of both flakes in equal proportion and the voltage of their junction remained the same. When the shutter opened, incoming radiation would be focused on the active element. The compensating element was shielded from external radiation, and because of this the active element experienced a change in temperature and its resistance became different to that of the compensating element. As a result there was a voltage change at the juncture of the two flakes and this change was connected to the electronics system.

As the scanning prism turned and caused the detector field of view to cross the horizon there was a sharp change in the radiation level striking the detector. The result of the radiation change during a complete scan cycle was the generation of an approximate square-wave signal at a frequency of 30 cycles per second.

The electronics system was physically arranged so that the functionally related parts were close to each other. It was divided into eight major circuits located on individual printed boards. The driving imperative to make the system as compact as possible meant that in some cases two or three sub-circuits were located on one board. Thus the functionally related booster amplifier, signal centring circuit and phase inverter-limiter were mounted on one board. Although the power supply and reference generator circuits were not closely related in function they were both located on the same printed circuit board.

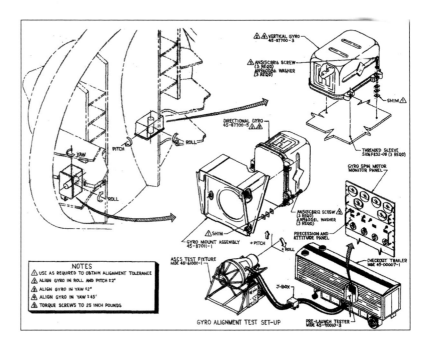

Gyroscopes

The purpose of the attitude gyros, both vertical and directional, was to determine the spacecraft's attitude angles between a fixed set of axes and the reference axis fixed in the orbital plane but moving with respect to the local vertical. Both attitude gyros were free in that they had slave capability. They also had a means incorporated for caging and for obtaining electrical signal (synchronisation) outputs that defined the attitude of the gyros with respect to two mutually perpendicular axes. The attitude gyros possessed unrestricted mechanical freedom in the outer axis and a minimum ±83° of mechanical freedom in the inner axis.

The rate gyros performed electrical switching functions at specific rates of angular velocity and an axis perpendicular to the base of each unit. This was referred to as the input axis. Rate gyros were used in the pitch, roll and yaw axes respectively, and each gyro consisted of a high-speed motor mounted in a gimbal ring. This was configured in such a manner that it was free to process about one axis only (referred to as the output axis), which was perpendicular to the spin axis of the rotor. The output signals were generated by the motion of the wipers attached to the gimbal ring, moving across the contacts of sector switches. Input power requirement for the rate gyros was 115V at 400 cycles per second.

ABOVE Calibration after installation of the attitude gyroscopes called for a checkout trailer hooked up to a J-box computer for aligning roll, pitch and yaw axes. The test set-up was configured during final assembly of the spacecraft. *(McDonnell)*

RIGHT The pitch axis block diagram identifies the logic routes for control and thruster authority commands, with delineation of the aspects not applicable to ASCS control. Note the added authority through instrument package 'B' for rate-output leverage and compare this to the roll and yaw axis diagrams. *(McDonnell)*

BELOW The roll axis diagram shows the control path terminating in the same torque-switching logic output relays as employed for the pitch axis. *(McDonnell)*

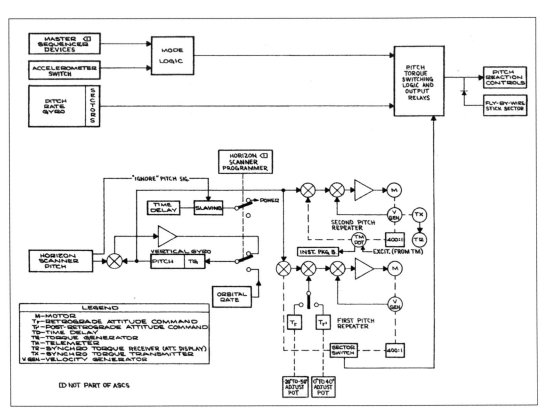

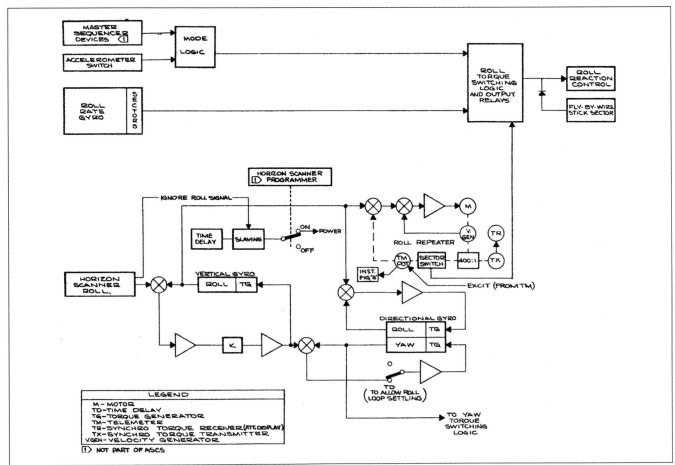

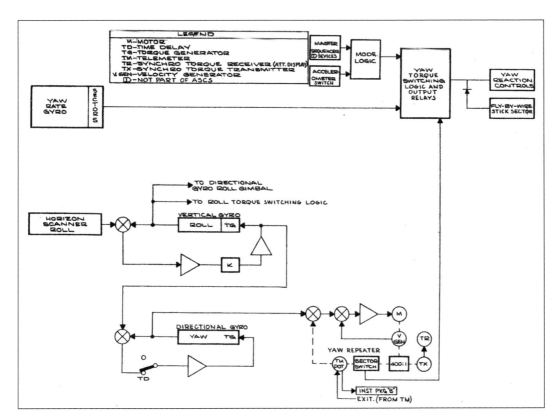

LEFT Yaw axis control goes through the same paths from sensor and gyroscope sources and instrumentation package 'B' as the roll and pitch control but it has a simplified path from the roll scanner and control logic. *(McDonnell)*

BELOW The manual reaction control system throttle valve and bellcrank operated by the torqueing handle on the astronaut's right-armrest location. See page 140 for the attitude control handle. *(McDonnell)*

The degree of gyro freedom did not necessarily reflect the attitudes permissible by manually steering the pointing angle of the capsule in orbit. Because of limitations in the horizon scanner and in the repeater section of the amplifier calibrator, manual control of the spacecraft was limited to ±30° in all three axes. Barring equipment malfunction, exceeding this rotational limit would not prejudice the operation of the equipment or the success of the mission. If the limits were exceeded, the manufacturer recommended that the gyro switch be placed in the 'FREE' position. The input requirements were 115V, 400 cycles per second, single phase for the gyro motor and 26V, 400 cycles per second for the synchronisation and torque motor.

The information from the gyroscopes was presented to the astronaut via the attitude and rate indicator mounted on the main instrument panel. This provided visual indications of capsule rate and attitude in pitch, roll and yaw planes. The attitude indicators were driven by the attitude gyro synchronisation outputs through the amplifier calibrator, which was functionally divided into four sections: slaving; repeating, mode switching and torque switching.

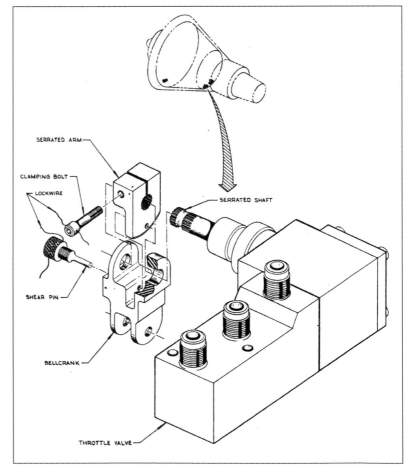

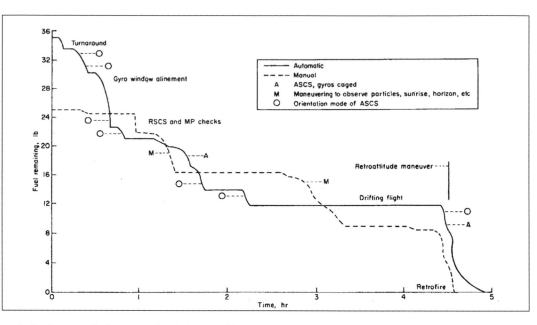

Attitude gyro slaving contained the amplifiers and summing networks that accepted roll and pitch information from the horizon scanners and generated current to torque units in the attitude gyros. Pitch and yaw gimbals were aligned with corresponding directions in, or perpendicular to, the orbit plane.

The repeater section was a group of servo-mechanisms. Attitude gyro inputs, received at the calibrator in proportional (or analogue) form, were amplified and then used to drive shafts. These served as roll, pitch and yaw signal sources for both internal (torque switching) and external (display and telemetry) purposes.

The mode switching section established the proper attitude angle bias, torque switching status and interlock signals corresponding to the ASCS mode commanded by the external devices. For the ASCS, external devices were summed to provide a master sequencer coordinating all automatic functions. The mode-switching section employed compact, solid-state switching circuits containing transistors, diodes and other electrical components of a class not completely dependent upon reference voltage or temperature levels.

The torque switching section contained the same array of electrical and electronic components as those installed in the mode switching section and received step-function outputs from the attitude gyro repeaters as well as the outputs of the rate gyros. Using the step-wise indications of attitude and rate conditions together with the mode switching section, defining the specific phase of the mission, allowed the astronaut to make decisions on

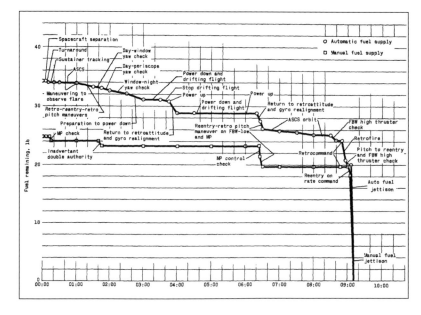

LEFT Tasked with a minimum-fuel usage flight plan, where emphasis returned to a priority for managing the spacecraft and conserving consumables in preparation for the one-day mission of Gordon Cooper, Walter Schirra flew MA-8 as a textbook demonstration of systems management. This consumption chart shows fuel-use rates throughout the six-orbit mission, resulting in a surplus after re-entry and prior to jettisoning before splashdown. *(NASA)*

selectively energising the appropriate RCS valves.

The astronaut's attitude and rate indicator was calibrated to indicate spacecraft attitude within a range of ±180°, except for yaw values which indicated 0°, 80°, and 270° in a clockwise direction. The rate portion of the indicator was driven by the miniature rate transducers that also served as sensing elements for the Rate Stabilization Control System. The range of rate indication was 0° to ±6°/sec for all three indicators. The roll rate indicator had the additional capability of being externally switched to a range of 0° to 15°/sec to enable the astronaut to monitor re-entry roll rate.

The miniature rate transducers consisted of a gyroscope, amplifier and a demodulator which together functioned to produce an AC output signal proportional to input rate-of-change attitude. All three rate transducers were identical except for gyro orientation in the transducer base. A special indexing feature prevented installation in the wrong location. Input power utilised by the rate transducers was 115V at 400 cycles/sec alternating current.

Gyro-related Rate Stabilization Control System activity involved the use of a rate damper box, which provided three channels of transistorised electronic units comprising the rate-damper portion of the RSCS. Each channel contained a summation network, preamplifier and demodulator and two trigger circuits. The box was small, 5.5in x 6in x 9in (14cm x 15cm x 23cm), weighing approximately 7lb (3.17kg), and was mounted below and immediately aft of the pilot's control stick. The rate damper box had two AN-type electrical connectors, one for the ground servicing equipment and one for capsule interconnection. The dead-band adjustments were 2°/sec to 4°/sec in pitch and yaw axes and 1.5°/sec to 3°/sec in the roll axis.

Signals from the rate transducers and stick potentiometers were sent to the rate damper box where the two were summed together to form an error signal, sent to a two-stage voltage amplifier and then to a double ring diode demodulator. Depending on the magnitude and phase of the error signals the demodulator would select one of two transistor-operated relays which applied a 24V DC output signal to appropriate solenoid control valves so that corrective torque could be applied.

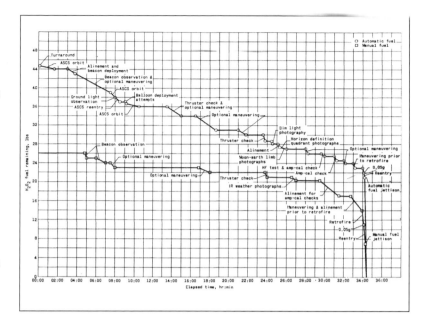

ABOVE Gordon Cooper's 22-orbit flight of MA-9 brought the flight programme to a close, with the astronaut demonstrating judicious use of both automatic and manual fuel supplies, as evidenced by the consumption trace over time displayed on this chart. It also demonstrated the trade-off choices between the use of automatic and manual systems. This allowed important choices to be made for the balance between computer-controlled and manual inputs on the Gemini spacecraft that followed Mercury. *(NASA)*

BELOW The Mercury spacecraft was designed to evaluate the performance of humans in space, not only their physiological and psychological reactions to weightlessness but also the impact of this environment on their performance as test pilots. The attitude control systems, vital for accurate pointing angles essential for retrofire and safe re-entry, pioneered a new set of equipment balancing automatic and manual authority. *(Courtesy of Bassett Celestia)*

Rocket systems

The Mercury spacecraft required a range of propulsion systems to separate it from the launch vehicle, control its attitude in orbit and bring it back to Earth. It also needed rocket motors to carry its lone occupant to safety in the event of a malfunctioning launch vehicle.

Reaction Control System (RCS)

The RCS consisted of two separate and independent automatic and manual systems, each providing control of pitch, roll and yaw axes using a pressure-fed, monopropellant/catalyst bed design, incorporating right-angle firing exhaust nozzles. The nozzles produced thrust through the decomposition of hydrogen peroxide (H_2O_2). The thrust chambers consisted of a stainless steel jacket with a metering orifice, a distribution disc followed by a catalyst bed and a nozzle. The catalyst bed contained a stack of removable nickel screen wafers and the screen gauge resembled that of a common household screen. The screen was covered with an electrolytically deposited coating of 99% silver and 1% gold (drexite) to enhance the catalytic properties of the nickel. The open area between the catalyst bed and the right-angle nozzle formed a short plenum chamber to smooth the flow prior to reaching the nozzle throat.

Hydrogen peroxide entered the thrust chamber through a metering orifice on activation of the solenoid valve and the stainless steel porous plate distributed the flow and presented the catalyst bed with a uniform layout. On entering the first stage of the catalyst bed a violent reaction took place. Expanding gases rushed through the remainder of the bed resulting in a peak thrust output in the right-hand nozzle. The majority of the decomposition, and the most violent, would occur within the first two catalyst cups where temperatures of 1,400°F (760°C) would be experienced. The remaining cups in the catalyst bed were there to ensure complete decomposition and to prevent any liquid form of hydrogen peroxide reaching the nozzle.

Hydrogen peroxide in any form is toxic and volatile. It is a clear and colourless liquid, soluble in all proportions in water and in most substances that are miscible with water. When decomposed, hydrogen peroxide releases water vapour, oxygen, gas and heat and when

OPPOSITE Apart from the attitude control thrusters, all the rocket systems on Mercury were there for safely securing the spacecraft in orbit, recovering the spacecraft from the end of its mission and providing a means of escape should anything go wrong with the launch vehicle. Atlas was chosen to place Mercury in orbit but it had a patchy success record. This Atlas 51D launched in March 1960 failed when combustion instability in a booster engine doomed it seconds after lift-off. It was for situations such as this that a launch escape system was essential. *(USAF)*

BELOW A page from the Little Joe-5A preflight familiarisation manual displays all the essential rocket systems carried by the Mercury spacecraft.
(McDonnell)

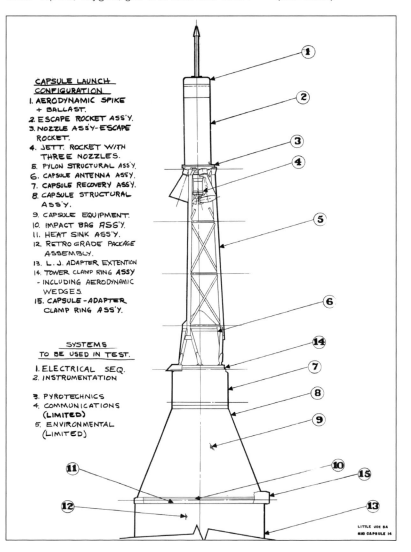

1972 1138 009

the decomposition is properly contained the resultant release of energy can, as in the case of the spacecraft's thrusters, produce workable energy for creating a reaction in the attitude orientation of the capsule. When decomposed in the most efficient manner, 1lb (0.45kg) of hydrogen peroxide will produce 60ft^3 (1.698m^3) of gas. H_2O_2 freezes at 11.3°F (-11.5°C) and boils at 286°F (141°C).

High-pressure helium at 2,250lb/in^2 (15,513kPa) was used to pressurise the hydrogen peroxide torus tanks, which passed through a filter and a manual shut-off valve to a regulator to reduce the pressure to 450lb/in^{2}g (3,102kPag). From there it passed to a check valve and finally to surround and apply pressure to the flexible bladder inside the rigid torus tank containing the fuel. The helium pressure forced the hydrogen peroxide out of the bladder in which it was contained and through a perforated tube downstream into the lines and valves. This form of positive expulsion was already becoming standard in lunar and planetary vehicles.

The manual push-pull shut-off valves that

allowed the hydrogen peroxide to be available at the solenoid valves provided a simple means of isolating and shutting off individual thrusters. When the ASCS sent a 24V DC signal the appropriate solenoid valve would open and hydrogen peroxide would pass to the corresponding thrust chamber where it was decomposed providing thrust from the exhaust nozzles.

For the automatic ASCS system, four nozzles delivered a thrust of 24lb (106.7N) for pitch and yaw axes and two nozzles delivered a thrust of 6lb (26.68N) for the roll axis. Two nozzles delivering 1lb (4.448N) were provided for each of the roll, pitch and yaw axes. For the manual RSCS system, two thrusters were assigned to each pitch and yaw axis and delivered rate-variable thrust levels of 4–24lb (17.8–106.7N). The roll axis had two thrusters with variable output of 1–6lb (4.44–26.68N). The manual control system had proportional control or rate command thrust levels between pitch and yaw axes, and these thrust levels were controlled through direct stick action by the astronaut or through the RSCS electronics.

The spherical helium tanks for the RCS thrusters were located in the cabin and fabricated from spherical fibreglass structure. Of half-toroidal shape each, they were contoured in that form to fit on the large pressure bulkhead in the cavity between the bulkhead proper and the heat shield. They were situated outside the pressurised area occupied by the astronaut because of the nature of the hydrogen peroxide. The tanks were of aluminium construction and insulated for proper temperature control. The tanks also were equipped with jettison vents activated by the astronaut following main

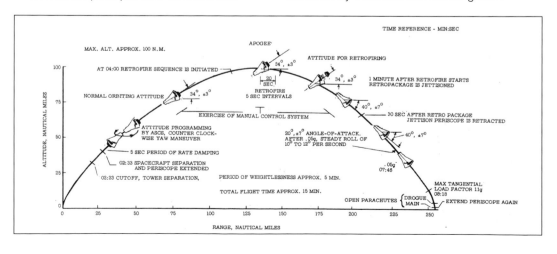

parachute deployment so that as much as possible of the remaining fuel could be removed prior to splashdown. Again, this precaution was for safety and to limit the release of toxic vapours to the astronaut emerging from the capsule or the recovery team engaged in retrieving man and spacecraft.

The two systems (ASCS and manual RSCS) operated in different ways. Each half-torus-shaped tank had a capacity for 32lb (14.5kg) of hydrogen peroxide and each spherical helium tank had a capacity of 265in³ (4,343cm³).

In ASCS operation the helium regulator manual shut-off valve was opened allowing the pressurant to pass through the filter, regular check valve and finally the bladder, pressurising it to 480lb/in² (3,310kPa). By opening the manual push-pull shut-off valve the H_2O_2 became available at the electrically operated solenoid, and upon receiving a 24V DC signal from the ASCS or FBW control system the appropriate solenoid valve would open.

Hydrogen peroxide would enter the solenoid valve through an integral 165-micron filtration screen and pass into the corresponding thrust chamber, where it would be decomposed to produce the desired thrust. The helium pressure transducer provided a means of monitoring the percentage of H_2O_2 present in the bladder and the perforated tube prevented the build-up of trapped helium while servicing the bladder.

The RSCS operation involved selection of either manually controlling proportional control valves or utilising electrically operated solenoid control valves. The only significant difference in the half-torus that served this system was that hydrogen peroxide capacity was 23.4lb (10.6kg). A two-position selector valve was provided which enabled the method of control to be selected. The manual control valves had a dead-band of ±0.0625in (0.165cm) from theoretical neutral and a total stroke of 0.37in (0.94cm) from theoretical neutral for each thrust chamber.

The throttle valve arm assemblies or bellcranks that rotated these proportional control valves were designed to shear at less than the full effort an astronaut could impose on the manual control system. If a proportional control valve should jam and immobilise the manual control system in one axis, increased hand-effort would break the shear pin in the bellcrank and

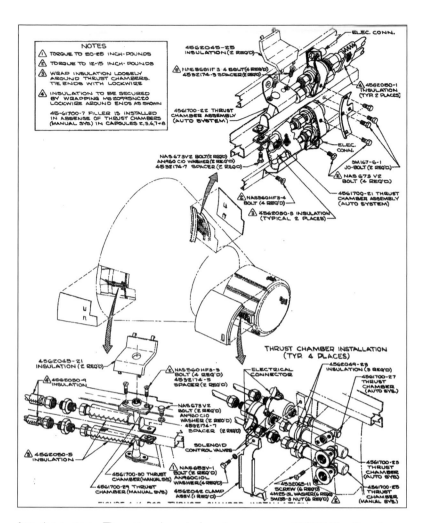

free the system. The manual control system could then be used for automatic control system fly-by-wire switch operation or manual control rate-stick potentiometer deflection.

Posigrade rocket system

Separation of the Mercury spacecraft from the launch vehicle adapter was aided by the firing of three posigrade rockets. The word 'posigrade' derives from a description of an auxiliary rocket motor designed to add thrust to the direction of flight, for instance to add an additional acceleration to increase the distance to a structure from which it has just separated – which was the precise function of those rockets on the Mercury spacecraft. They were important so as to prevent the separated stage of the launch vehicle from colliding with the spacecraft and potentially causing damage. Even after engine shutdown with the launch vehicle there was still latent inertia from thrust tail-off when

ABOVE The thrusters were located at the apex of the conical spacecraft around the periphery of the recovery compartment and at the base end within the space between the main bulkhead and the heat shield, as shown here. *(McDonnell)*

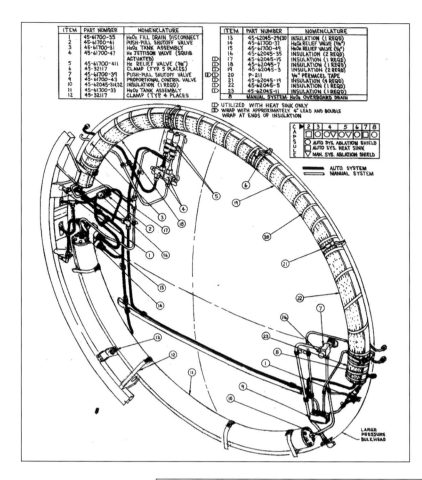

ITEM	PART NUMBER	NOMENCLATURE		ITEM	PART NUMBER	NOMENCLATURE
1	45-61700-35	H₂O₂ FILL DRAIN DISCONNECT		13	45-62045-29(30)	INSULATION (1 REQD)
2	45-61700-41	PUSH-PULL SHUTOFF VALVE		14	45-61700-31	H₂O₂ RELIEF VALVE (¾")
3	45-61700-31	H₂O₂ TANK ASSEMBLY		15	45-61700-49	H₂O₂ RELIEF VALVE (⅛")
4	45-61700-47	He JETTISON VALVE (SQUIB ACTUATED)		16	45-62045-35	INSULATION (2 REQD)
5	45-61700-411	He RELIEF VALVE (⅜")		17	45-62045-15	INSULATION (1 REQD)
6	45-32117	CLAMP (TYP. 5 PLACES)		18	45-62045-7	INSULATION (1 REQD)
7	45-61700-39	PUSH-PULL SHUTOFF VALVE		19	45-62045-3	INSULATION (2 REQD)
8	45-61700-43	PROPORTIONAL CONTROL VALVE		20	P-211	¾" PERMACEL TAPE
10	45-62045-31(32)	INSULATION (1 REQD)		21	45-62045-19	INSULATION (4 REQD)
11	45-61700-33	H₂O₂ TANK ASSEMBLY		22	45-62045-19	INSULATION (1 REQD)
12	45-32117	CLAMP (TYP 4 PLACES)		23	45-62045-11	INSULATION (1 REQD)
				8		MANUAL SYSTEM: H₂O₂ OVERBOARD DRAIN

▷ UTILIZED WITH HEAT SINK ONLY
▷ WRAP WITH APPROXIMATELY 4' LEAD AND DOUBLE WRAP AT ENDS OF INSULATION

AUTO. SYSTEM
MANUAL SYSTEM

LARGE PRESSURE BULKHEAD

residual exhaust products were escaping from the rocket's nozzle. That would give the stage added motion while the spacecraft was inert and at a fixed velocity – which was why the posigrade system was so important for the safety of the spacecraft.

Provided by Atlantic Research Corporation, the three posigrade rockets on the Mercury spacecraft were mounted symmetrically within the retro-package assembly together with the associated electrical wiring to ignite the rockets at the required time. The rockets were designed to impart a separation rate of 15ft/sec (4.6m/sec) or about 10mph (16.1kph) and to fire simultaneously. Any two were sufficient to provide an acceptable separation velocity, however, and the same was true when they were used to support an abort off the top of a launch vehicle after tower jettison.

The rockets were essentially a nozzle assembly and case, a solid propellant and an electrically actuated igniter. Each rocket was cylindrical in shape with a length of 14.7in (37.3cm) and a diameter of 2.8in (7.1cm). Each weighed about 5.24lb (2.37kg) and contained Arcite 377 (a compound based on asphalt),

ABOVE Installed at the base, the aft thrusters were located close to the toroidal fuel tanks containing hydrogen peroxide. This page from the Mercury maintenance manual identifies the location for spacecraft 2 to 8. *(McDonnell)*

RIGHT Location and installation equipment associated with the RCS helium pressurisation system for the automatic (shaded) and manual systems. *(McDonnell)*

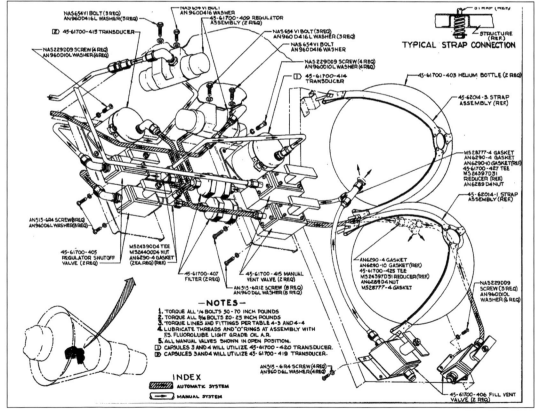

—NOTES—
1. TORQUE ALL ¼ BOLTS 50 - 70 INCH POUNDS
2. TORQUE ALL ³⁄₁₆ BOLTS 20 - 25 INCH POUNDS
3. TORQUE LINES AND FITTINGS PER TABLE 4-3 AND 4-4
4. LUBRICATE THREADS AND "O" RINGS AT ASSEMBLY WITH FS. FLUOROLUBE LIGHT GRADE OIL A.R.
5. ALL MANUAL VALVES SHOWN IN OPEN POSITION.
▷ CAPSULES 3 AND 4 WILL UTILIZE 45-61700-420 TRANSDUCER.
▷ CAPSULES 3 AND 4 WILL UTILIZE 45-61700-418 TRANSDUCER.

INDEX
AUTOMATIC SYSTEM
MANUAL SYSTEM

RIGHT Some way of quickly increasing the separation distance between the spacecraft and the launch vehicle was necessary after the mechanical connection was severed. To prevent Atlas from shunting into a warhead in its primary role as a ballistic missile, it was fitted with retro-rockets firing forward. To prevent exhaust impingement on the Mercury spacecraft's heat shield, three posigrade rockets were instead installed within the retro-package attached to the base of the capsule, pushing it forward from the inert Atlas. *(McDonnell)*

which produced a thrust of 370lb (1,645N) for one second. The igniter was mounted in the head of the cylinder and had dual ignition capability. Cylindrical in shape, it had a hexagonal head for threading it into the top of the posigrade rocket and packed about 3gm of ignition pellets ignited by either of two pairs of squibs. Each pair had independent circuitry from separate power sources and either one was capable of igniting the pellets.

Activation of the rockets was accomplished by a signal from the capsule-adapter ring separation sensor after firing of the clamp ring explosive bolts. The integrating accelerometer, energised by the ring separation sensor, measured the velocity increase and also initiated onboard recording and telemetry. This action was transmitted to the receiving stations on Earth but a delay relay cut off the signal-to-ground after five seconds.

Retrograde rocket system

Re-entry into the Earth's atmosphere was effected by a retrograde rocket system consisting of three solid propellant rocket motors together with associated components and wiring. The orbit for the Mercury spacecraft was of sufficiently low altitude as

RIGHT The retrograde package contained the three retro-rockets and the three small posigrade rockets within a drum-shaped structure held to the base of the heat shield by three metal straps.

(McDonnell)

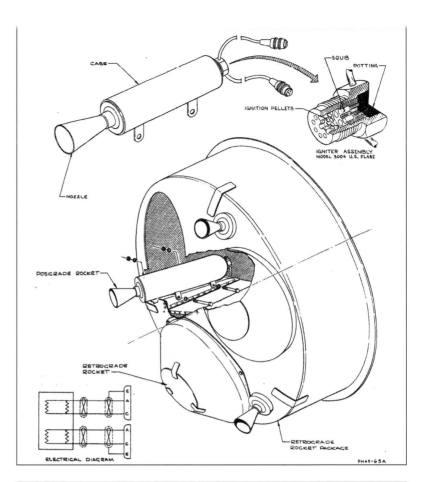

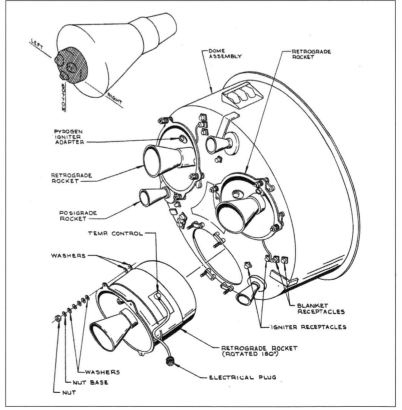

ABOVE A full-scale retro-rocket package with retro-rockets and posigrade rockets, the latter inside a protective cover during launch. *(NASA)*

ABOVE RIGHT Producing an average thrust of almost 1,000lb for 10–11 seconds, the retro-motors produced by Thiokol were adapted from the same motors used to de-orbit the Discoverer capsules used in America's first spy satellite programme. *(Thiokol)*

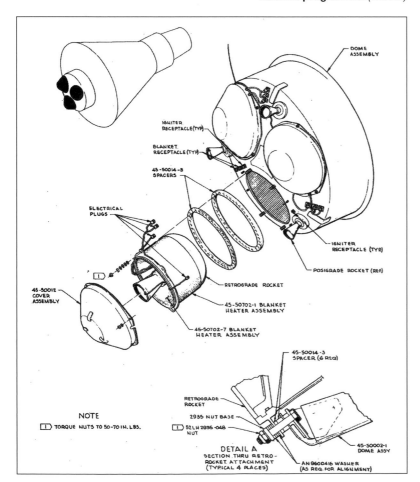

NOTE
[1] TORQUE NUTS TO 50-70 IN. LBS.

DETAIL A
SECTION THRU RETRO-
ROCKET ATTACHMENT
(TYPICAL 4 PLACES)

to allow natural decay back down through the atmosphere for operational reasons, and because of that the velocity decrement required of the retrograde system was a mere 500ft/sec (152m/sec), or 341mph (549kph). The design objective was to provide a total vacuum impulse of about 13,000lb/sec (5,897kg/sec) providing an average thrust of 992lb (4,414N) for an action time of 13.2 seconds.

The retrograde rocket assembly was manufactured by Thiokol Chemical Corporation. It had a weight of 69.55lb (31.5kg) and a total impulse of approximately 13,000lb/sec (5,898kg/sec) providing an average thrust of 992lb (4,412N) each for an action time of about 13.2 seconds. The rockets were equipped with blanket-style heaters operated by ground power supplies before launch. A resistor-type thermostat maintained temperatures between 75–95°F (24–35°C), ± 5°F, to prevent the build-up of moisture on the nozzle enclosures. The thermal blankets were used to keep the motors warm during assembly and also helped maintain even temperatures in orbit. The 12in (30.5cm) diameter, 15.5in (39.4cm) long motor had a concave bulkhead to reduce the length of the nozzle that extended below the structure. A fibreglass-plastic exit nozzle saved weight.

LEFT The retro-rockets were held securely in place with spacers and fixtures incorporating thermal blankets and a cover blown off at ignition. *(McDonnell)*

Reliability was of supreme importance since these were the only motors carried by the spacecraft capable of bringing the capsule down into the atmosphere. To enhance reliability, twin pyrogens – tiny solid charges – were employed to ignite the solid propellant in each motor, providing low-pressure ignition. Because the reliability of the Thiokol retro-rockets was an essential element in their design almost 190 motors were tested at the company's Elkton, Maryland, plant and in the wind tunnel at Arnold Engineering Development Center, Tullahoma, Tennessee. The design of the motor was essentially that of the retro-rocket used to bring capsules back to Earth in the Discoverer series of satellites, the majority of which served in the first-generation spy satellite programme.

The retro-rockets were mounted in the retro-package assembly, each aligned so that the line of thrust went approximately through the centre of gravity of the spacecraft. This alignment was checked and verified on the pad and before flight. The assembly was secured to the capsule with three straps, each of which carried an explosive bolt. Sixty seconds after the retrograde firing signal this bolt detonated and released the straps, a coiled spring pushing the package away from the base of the heat shield. Each rocket had a metal cover to protect it from micrometeorite impacts and this cover was blown off at ignition.

Electrical power was provided through the three explosive disconnect points equally positioned around the base of the capsule at 120° intervals. Wide bundles from the disconnect points followed the three retrograde package retaining straps down to the retro package where they entered the unit through rubber grommets. Within the package, all the wiring for the retro-rockets, sensors and heaters was routed to the exterior face and then on to each retro-rocket through the slotted metal shield. Wiring for the posigrade rockets and explosive bolt was retained within the package.

Firing of the retro-rockets followed the 30-second period during which the capsule was orientated in retro attitude at 34°, ±12.5°, in pitch and 0°, ±30°, in roll or yaw. The firing sequence was inhibited until the retro attitude had been achieved within the design limits and would be interrupted if the attitude of the spacecraft drifted out of those bands. These constraints could, however, be manually overridden in emergencies. Firing of the rockets occurred at five-second intervals and could be initiated either by the clock running down to activation, manual selection by the astronaut or by ground command. Any single rocket would effect a re-entry should the other two fail.

All three rocket fire relays received a 24V DC signal at the same time. However, the No 2 and No 3 rocket relays had a five- and ten-second delay respectively. In a normal sequence, the ignition of the first rocket was followed 5.5 seconds later by ignition of the second rocket, the two burning for a further 5 or 6 seconds until the first rocket burned out and the third rocket motor ignited, joining the second until it burned out halfway through the burn time of the third motor. In this way, the beginning and the end of the 22-second retro-fire period was covered by a single motor burn, relaxing the forces imposed on the spacecraft and allowing gradual build-up and tail-off to the process.

BELOW The retrograde package electrical systems required for sequential operation of respective posigrade and retrograde motors. Thermal blankets and small heaters kept the system warm relative to the freezing conditions of space. *(McDonnell)*

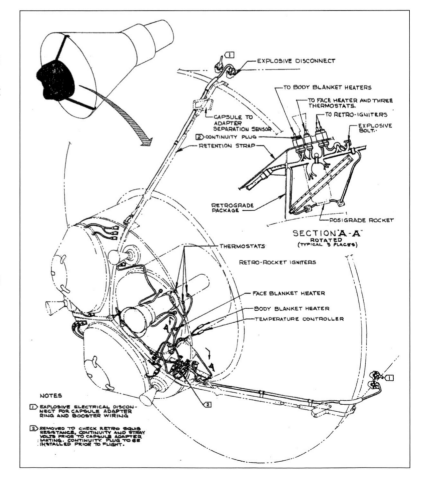

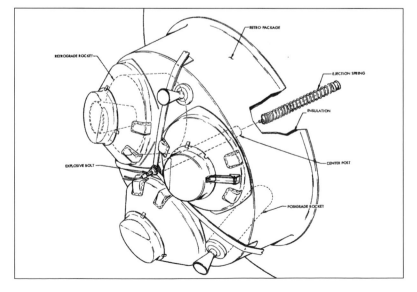

LEFT A retrograde package undergoes test at NASA's Lewis Research Center. *(NASA)*

CENTRE The test stand at Lewis Research Center was employed to test the separation pyrotechnics and to evaluate the force required to sever the retaining straps. *(NASA)*

Sequential firing was also designed to avoid the ineffective results of a failure with either of the first two firings. If the No 1 rocket failed, the ASCS attitude interlock would remove power from the attitude permission relay and also the No 2 rocket fire relay. The spacecraft would then reposition itself either automatically or by manual operation of the RCS, and upon regaining the 34° pitch-down position the No 2 relay would receive electrical power and fire. The same sequence would follow should a failure occur with the No 2 rocket.

The emergency override involved four telelights on the astronaut's left instrument panel. The first was 'RETRO SEQ' which had a green function illuminated when the retro sequence began, started either by the clock or the button adjacent to the light. The purpose of this button was to initiate re-entry before the capsule clock ran out or failed. The next two lights were the 'RETRO ATT' and 'FIRE RETRO'. During normal flight the 'RETRO ATTITUDE' switch adjacent to the 'RETRO ATT' switch would be in 'AUTO' and its associated light would glow green when the capsule reached the 34° retro orientation. About 40 seconds later the 'FIRE RETRO' light would glow green, but if it should show red the astronaut would be required to check attitude to verify that the capsule was appropriately

LEFT Designed to eject the retrograde package after the de-orbit manoeuvre, a coiled spring pushed the assembly away as the spacecraft remained in weightlessness and before deceleration could impose positive g-forces to hold it on. On John Glenn's MA-6 mission this package was retained all the way back through the atmosphere as great chunks of it burned away, giving him the impression that his spacecraft was burning up. See Chapter 4. *(McDonnell)*

orientated for ignition. If it was confirmed to be in the correct attitude the astronaut would position thc 'RETRO ATT' switch in the 'BYPASS' position and push the 'FIRE RETRO' button.

Ten seconds later the 'FIRE RETRO' would illuminate green, but if the astronaut found that the capsule was not in the correct position the fly-by-wire system would be used to correctly position the spacecraft for the burn. When this was accomplished the light would show green. If the 'RETRO ATT' light showed green but the 'FIRE RETRO' telelight shone red, the button adjacent to the 'FIRE RETRO' light would be pushed, leaving the 'RETRO ATT' switch in the 'AUTO' position.

The fourth telelight was the 'JET RETRO', which would illuminate green 60 seconds after the 'FIRE RETRO' telelight showed. But if it was red the adjacent button would be selected to supply an alternate source of power to the jettison bolt. If the retro package could not be jettisoned either automatically or by override method it would be removed during re-entry when the heat would detonate the explosive bolt or burn through the retention straps to allow the coiled spring to eject the package.

LEFT Operational spacecraft carried retrograde packages with black and white stripes for passive thermal control. Note the relative location of the small posigrade motors. *(Via David Baker)*

Escape rocket

The escape system for use during on-pad and ascent phases of a typical mission included an escape rocket motor mounted on a pylon, or tower, with a jettison rocket to carry the tower and assembly free, either when discarded at the scheduled time during a normal flight or after successfully carrying the spacecraft away from a malfunctioning launch vehicle. Next to the retro-rocket assembly, the escape rocket was probably the most important rocket system carried by the spacecraft, for without it there was no means of escape until

RIGHT A Jupiter rocket lifts off from LC-26B at Cape Canaveral. This rocket had been mooted as a possible test vehicle carrying a Mercury spacecraft to fill the performance gap between the Little Joe and the Atlas. It was rejected when costs proved it uneconomic, but the launch escape system was adequately tested by the other two rockets. *(US Army)*

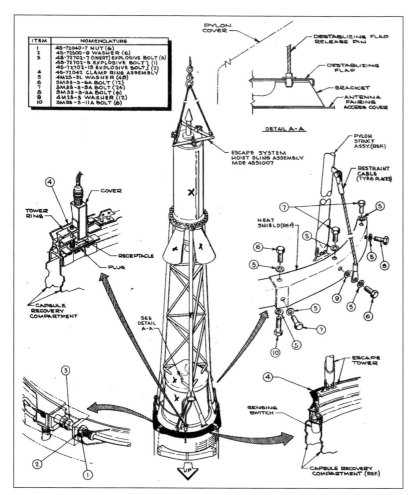

ITEM	NOMENCLATURE
1	45-72040-7 NUT (6)
2	45-72500-9 WASHER (6)
3	45-72702-7 (INERT) EXPLOSIVE BOLT (3)
	45-72702-5 EXPLOSIVE BOLT (1)
	45-72702-15 EXPLOSIVE BOLT (2)
4	45-72042 CLAMP RING ASSEMBLY
5	4M25-3L WASHER (48)
6	3M33-3-6A BOLT (12)
7	3M33-3-3A BOLT (24)
8	3M33-3-3A BOLT (6)
9	4M25-3 WASHER (12)
10	3M33-3-11A BOLT (6)

ABOVE The Mercury launch escape system was a weighty solution to a significant problem – the essential requirement for carrying the astronaut away from a launch vehicle in trouble. As hoped, it was never used with an astronaut aboard. *(McDonnell)*

LEFT Engineers prepare a Mercury boilerplate for launch atop a Little Joe, one of several tests to evaluate the performance of this escape system. *(NASA)*

the spacecraft reached such an altitude that it could itself separate from the launch vehicle using the three posigrade rockets.

The tower supporting the escape rocket consisted of three longitudinal members of tubular steel construction diagonally braced and attached to the capsule cylindrical recovery compartment. A 45° aerodynamic fairing was installed over the pylon clamp ring to reduce the pylon ballast weight and to give the structure greater aerodynamic stability up to tower separation. The clamp ring assembly consisted of three segmented sections each joined by explosive bolts, two of which were initiated electrically from one end and by a gas generator source at the other end through a percussion system.

The escape rocket had a length of 70in (178cm) and a diameter of 15in (38cm) with a launch weight of 350lb (159kg), and for added stability ballast was loaded in the top of the rocket. The main body of the rocket motor was fabricated from 0.25in (0.63cm) 4130 steel with a plenum chamber and a solid fuel propellant. The nozzle assembly incorporated three equally spaced exit cones, each canted 19° from the centreline so that exhaust plumes would not impinge on the top of the spacecraft. The plenum chamber included a jettison motor boss that also provided for the attachment of the thrust alignment mirror. The optical lighting of the resultant thrust vector was accomplished by this mirror. For aborts off the launch pad, the escape system would propel the spacecraft to an altitude of about 2,500ft (762m).

The propellant for the escape rocket was a polysulphide ammonium perchlorate formulation that was widely used in the rocket industry at the time, and has in fact remained so ever since the Mercury programme. The US Bureau of Explosives categorised the propellant as a Class B Explosive, defining it as sensitive to pressure and easily ignited by a spark or flame. The propellant grain incorporated an internal burning nine-point star that was cast directly into the case and bonded to its sides. Improvements to

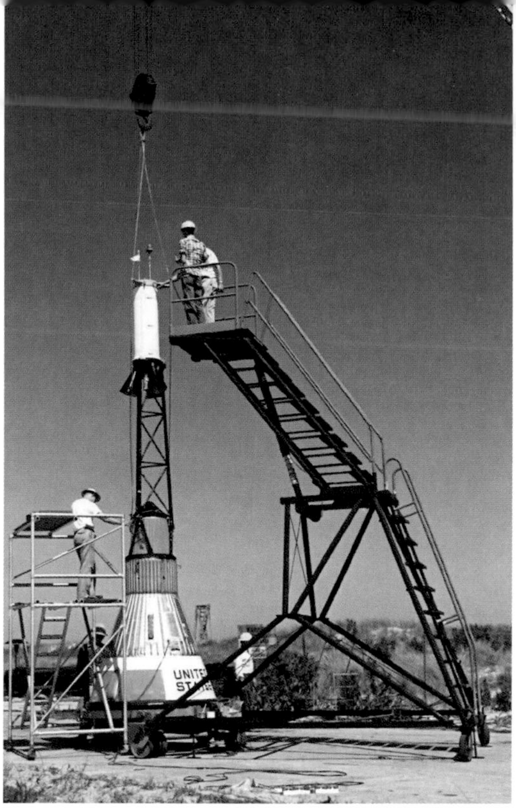

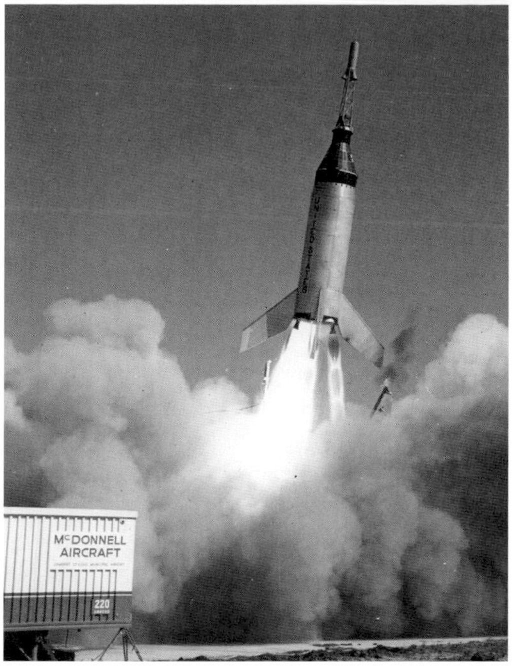

FAR LEFT **Preparations under way at Wallops Island for a Little Joe flight (LJ-6) in October 1959, in which a Mercury boilerplate would evaluate the aerodynamic performance of the capsule and its escape tower.** *(NASA)*

LEFT **Launched on 28 April 1961, Little Joe-5B tested the launch escape system when spacecraft 14A was put through a flight profile that satisfactorily simulated the worst conditions an Atlas rocket could impose on the capsule.** *(NASA)*

the nine-point star design greatly reduced the possibility of thrust vector due to a more accurate shaping alignment between the points of the star and the centrelines of each exhaust nozzle.

The igniter for the escape motor was a head-mounted dual ignition unit, each one separate from the other and with independent circuitry from different batteries. The igniter was cylindrical in shape and was of the central dynaflow type with a long burn from either of two squibs of a boron-potassium nitrate pellet. It was surrounded by an annular plastic tube filled with a metal-oxidant compound in which were situated the two sets of four squibs. The igniter was assessed as a Class A substance. The nominal axial thrust of this escape motor was 56,000lb (250.88kN) at 70°F (21°C) for 0.78 seconds, after which thrust tailed off to 5,000lb (22.24kN) during the next 0.6 seconds. Thrust then diminished to zero for a total impulse of 56,500lb/sec (251.3kN/sec).

During normal flight, where the escape system was not used, the escape rocket motor would be employed to separate the tower from the ascending stack of spacecraft and launch vehicle. In the event of an abort, however, where the escape rocket motor had fired to carry the capsule to safety, a separate jettison rocket was necessary to carry the spent escape rocket and tower assembly away from the freed spacecraft.

The jettison rocket was manufactured by Atlantic Research Corporation and was an existing and flight-qualified retro-rocket from the Thor missile programme. It consisted of an electrically actuated igniter, motor case and tri-nozzle assembly. The cylindrical jettison motor igniter was head-mounted with dual ignition capability. The attachment of the hexagonal cap was threaded into the top of the jettison rocket. The igniter itself contained approximately 0.2oz (7gm) of USF-2D ignition pellet. The thrust axis of each nozzle was canted at a 30° angle, as measured from the centreline of the motor case

BELOW **The launch escape motor was a single case of solid propellant discharging its exhaust through a triple-nozzle arrangement so as to deflect the hot gases away from the spacecraft beneath.** *(McDonnell)*

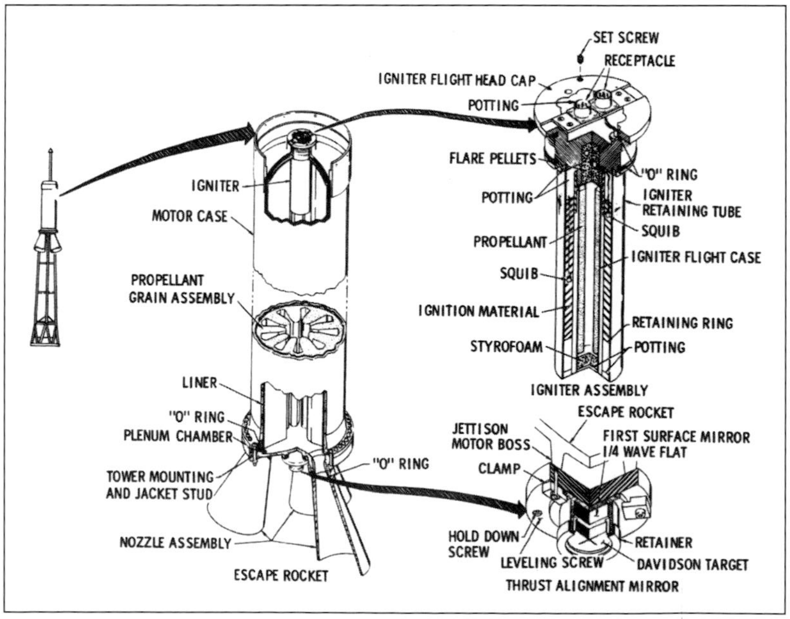

to the centreline of the nozzle. It weighed 19.5lb (8.8kg) and had a length of 18in (45.7cm) and a diameter of 5.5in (14cm). The motor had a thrust of 785lb (3.49kN) for one second and had been successfully rated at temperatures from -75°F (-59°C) to 175°F (79°C) from sea level to the vacuum conditions of space.

Manual activation of the escape system could be initiated by the pilot through an abort handle, used as a restraint handle during launch. The handle was located forward of the astronaut's support couch left armrest and to activate it the release button on top had first to be depressed, which allowed the handle to rotate outboard of its restrained position. When moved to the outboard position an electrical switch was automatically activated, which acted to detonate the clamp ring holding the capsule to the launch vehicle adapter. The escape sequence was then automatically activated providing that the main umbilical connecting the launch vehicle to the ground had been disconnected.

As related earlier, the escape motor propelled the spacecraft to an altitude of 2,500ft (762m) beyond the point of activation. Two seconds after tower jettison the drogue chute would be deployed, followed two seconds after that by release of the antenna fairing. Twelve seconds later the heat shield would be released, extending the landing impact bag skirt.

The escape system operated any time after

the removal of the launch gantry to the time of tower separation at 2min 34sec, 23 seconds after separation of the twin Atlas booster motors in the case of an orbital flight, or at booster cut-off during a Redstone suborbital flight, 2min 22sec after lift-off. In total the escape tower assembly had a weight of 1,073lb (487kg) which represented approximately 26% of the total weight lifted by the launch vehicle – on average 4,124lb (1,870kg) including the spacecraft, which weighed 3,945lb (1,789kg) including the adapter and the main clamping band.

Activation of the escape tower before disconnection of the umbilical connecting the spacecraft to the ground could only take place through a single controlled signal sent through a direct hard-line from the blockhouse abort switch through the rocket to the capsule's Mayday relays. However, if the capsule had to be fired away from the launch vehicle and the missile was unable to transmit the signal, umbilical pins 44 and 45 were abort-wired and could transmit 28V power from the blockhouse to the capsule ground command abort signal latching relay, energising and locking-in that relay.

Through this energised relay, 28V DC squib arm bus power would be transmitted to the pole of the ground-test umbilical relay. Power would not continue through this relay until it was de-energised and the only way to do that was to eject the umbilical. If this mode was chosen it would be necessary for the test conductor in the blockhouse to first select the abort switch delivering power to pins 44 and 45, which would be followed milliseconds later by umbilical ejection.

Depending on the precise sequence of pre-launch events the umbilical would be separated from the launch vehicle between 50 seconds and 90 seconds prior to launch. This wedge of time, up to the lift-off signal given when the launch vehicle had risen 2in (5cm) from the launch restraints, provided three separate methods of activating the escape sequence: blockhouse to rocket hard-line abort signal as outlined above; ground command receiver abort signal; and activation by the astronaut inside the capsule by means of the abort handle, also as described earlier.

For escape after lift-off and before scheduled tower separation, three methods were possible: ground command receiver abort signal; use of the astronaut's abort handle at the discretion of the occupant; or activation of the booster catastrophic failure detection system. Prior to lift-off the time-zero relay would be de-energised but at the 2in (5cm) lift-off signal the relay completed a circuit to the Mayday relays. The catastrophic failure detection relay was de-energised through loss of power from the launch vehicle.

If the Mayday relays were energised in the event of an in-flight abort, the abort light would illuminate and the abort sequence would commence, which included shutting down the launch vehicle propulsion system. The thrust sensor would detect a decay of 0.20g in acceleration, which would apply power to the capsule's one-second time delay relay after which the separation bolts' power relay together with the capsule ring interlock relay would be energised. Power would also flow to the capsule separation warning light time-delay relay and the tower jettison 15-second time-delay relay.

After those 15 seconds the abort relay in the maximum altitude sensor would be energised and this would compute the time required for the spacecraft to reach a safe dynamic pressure before jettisoning the escape tower. The capsule separation bolts would fire, releasing the capsule adapter clamp ring and allowing the three limit switches to return to their normal positions, energising the emergency escape rocket fire relay and the capsule adapter disconnect squib fire relay. This would fire the escape rocket and the four capsule adapter explosive disconnect squibs, releasing the spacecraft and tower to ascend away from, and slightly to one side of, the failing launch vehicle.

Capsule separation sensor limit switches would also energise the capsule separation sensor relays that turned on the green capsule separation telelight and simultaneously energised the tower separation abort interlock latching relay. This in turn energised the retro-rocket assembly jettison relay through the 'ARM' position of the astronaut's 'AUTO RETRO' switch and fired the two squibs of the retro-rocket assembly jettison bolt. When this bolt fractured the retro-package would drop free of the heat shield, assisted by the coil spring located between the shield and the pack.

When the capsule reached maximum altitude, contacts in the 'MAX ALT SENSOR' would close and energise the tower separation bolts' power

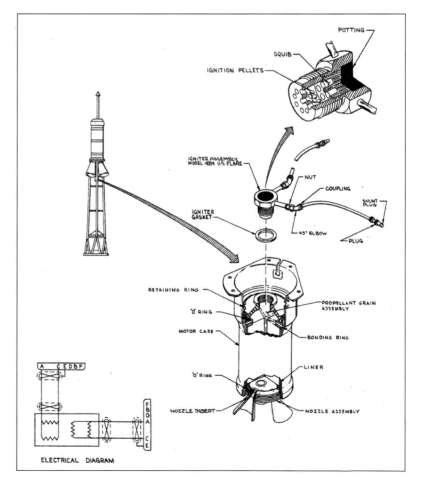

BELOW The tower jettison motor nested within the escape rocket, which never failed on a single test and was integral with the escape motor casing. (McDonnell)

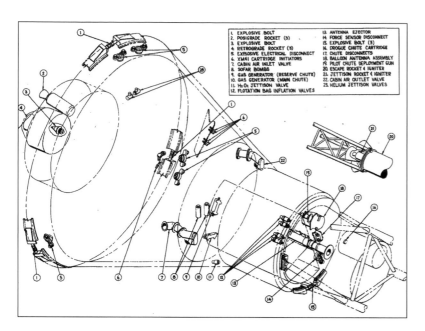

1. EXPLOSIVE BOLT	13. ANTENNA EJECTOR
2. POSIGRADE ROCKET (3)	14. FORCE SENSOR DISCONNECT
3. EXPLOSIVE BOLT	15. EXPLOSIVE BOLT (3)
4. RETROGRADE ROCKET (3)	16. DROGUE CHUTE CARTRIDGE
5. EXPLOSIVE ELECTRICAL DISCONNECT	17. CHUTE DISCONNECTS
6. XM41 CARTRIDGE INITIATORS	18. BALLOON ANTENNA ASSEMBLY
7. CABIN AIR INLET VALVE	19. PILOT CHUTE DEPLOYMENT GUN
8. SOFAR BOMBS	20. ESCAPE ROCKET & IGNITER
9. GAS GENERATOR (RESERVE CHUTE)	21. JETTISON ROCKET & IGNITER
10. GAS GENERATOR (MAIN CHUTE)	22. CABIN AIR OUTLET VALVE
11. H₂O₂ JETTISON VALVE	23. HELIUM JETTISON VALVES
12. FLOTATION BAG INFLATION VALVES	

ABOVE The spacecraft carried many pyrotechnic explosive bolts and charges for severing various elements and separating redundant pieces of equipment, as shown on this layout diagram.
(McDonnell)

OPPOSITE Alan B. Shepard is hoisted from his spacecraft after splashdown following the first American manned space flight on 5 May 1961, the success of which validated President Kennedy's plan to send men to the Moon, announced less than three weeks later – arguably the greatest success of the Mercury project.
(NASA)

relay, firing the three bolts holding the tower to the top of the spacecraft. Simultaneously the segments' tower ring clamp would separate, allowing the three tower ring limit switches to return to their normal position, energising the emergency jettison and jettisoning the rocket fire relays. Through these relays and their parallel contacts the main and isolated bus power would fire the two squibs of the jettison rocket.

In the process of separating from the spacecraft the two capsule-to-tower electrical disconnects would energise the tower separation sequence sensor relays, powering in turn the abort rate damping relay. This action would send a signal to the ASCS commanding rate damping until the main parachute deployed. De-energising the tower separation relays would also start the two two-second timers in the recovery sequence which would arm the 21,000ft (6,400m) and 10,000ft (3,048m) baro-switches after two seconds.

Recovery equipment

The landing and recovery equipment provided a fully automated descent and splashdown sequence after an abort or re-entry, and the primary equipment to achieve that included a drogue parachute, a main parachute and a reserve parachute. After impact the parachute would automatically disconnect and the reserve parachute would be ejected. The astronaut was then free to leave the spacecraft

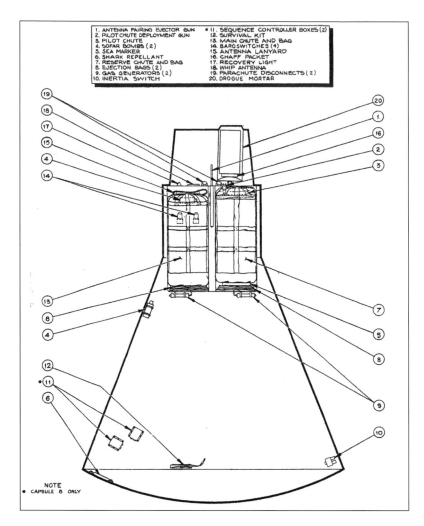

Key to diagram:

1. ANTENNA FAIRING EJECTOR GUN
2. PILOT CHUTE DEPLOYMENT GUN
3. PILOT CHUTE
4. SOFAR BOMBS (2)
5. SEA MARKER
6. SHARK REPELLANT
7. RESERVE CHUTE AND BAG
8. EJECTION BAGS (2)
9. GAS GENERATORS (2)
10. INERTIA SWITCH
*11. SEQUENCE CONTROLLER BOXES (2)
12. SURVIVAL KIT
13. MAIN CHUTE AND BAG
14. BAROSWITCHES (4)
15. ANTENNA LANYARD
16. CHAFF PACKET
17. RECOVERY LIGHT
18. WHIP ANTENNA
19. PARACHUTE DISCONNECTS (2)
20. DROGUE MORTAR

NOTE
* CAPSULE 8 ONLY

ABOVE The arrangement of parachutes and associated equipment in the spacecraft's recovery compartment, including survival package and various items used after splashdown, identified here and described in the text. *(McDonnell)*

RIGHT Egress training shows the ungainly manner in which an astronaut was likely to emerge from his spacecraft after splashdown. *(NASA)*

either by working his way to the top of the capsule and exiting through the hatch on the upper bulkhead or via the explosively released side hatch in extreme emergency.

Parachutes

The drogue parachute assembly consisted of a conical ribbon-type canopy with integral riser, deployment bag, mortar, sabot, chaff packet and drogue mortar cover. The conical ribbon drogue had eight gores of 2in (5cm) wide, 460lb (209kg) tensile strength ribbons and eight tubular suspension lines of 1,000lb (456kg) tensile strength each. The parachute constructs itself to a diameter of 6.85ft (2.09m) and was permanently reefed to an effective diameter of 6ft (1.83m) by means of pocket bands. The constructed total porosity was 27.9% and the effective porosity through reefing was 36.3%.

The 30ft (9.1m) integral riser was made from three layers of 3,000lb (1,361kg) tensile strength low-elongation, hot-stretched Dacron webbing. The function of the drogue was to stabilise and decelerate the capsule's rate of descent. It weighed 2.9lb (1.31kg) without riser and 5.9lb (2.67kg) with Dacron riser included. The deployment bag served to protect the parachute during ejection from its container and to provide a smooth and orderly deployment.

The bag containing the drogue parachute was manufactured from cotton sateen fabric reinforced with nylon webbing and covered at the upper end with a heat insulator of glass cloth. The bag was weighted at its upper end with a 0.5lb (0.23kg) lead disc that assisted in stripping the bag from the canopy after the line and riser had stretched out. Inside the bag were four cotton tapes to which the riser was secured during packing so that this could assist with a smooth deployment. The mouth of the bag was closed with a light cotton tie cord.

The drogue parachute mortar and sabot was a device designed to eject the drogue parachute with sufficient energy to overcome the local pressure gradients and gravitational forces. Inside its bag, the drogue parachute was stowed in the mortar tube on top of a lightweight sabot. The sabot operated effectively as a piston to eject the parachute pack when pressurised by gas generated from a pyrotechnic charge. This charge was initially fired into a breech chamber of small volume to produce high-pressure gas which subsequently vented through a small orifice and into the main chamber at relatively low pressure. In this way reaction loads were kept to a minimum because the pressure energy was not expended instantaneously.

The pressure sealing quality of the sabot was derived from an O-ring installed within a groove near the base. Two small holes were located in the groove to vent air trapped in the mortar tube beneath the sabot on installation. For proper operation the O-ring and the inner wall of the mortar tube were lubricated. The drogue parachute pack was retained in the stowed position within the mortar tube by a thin metal cover fabricated from René-41 that was attached to the upper surface of the antenna housing on top of the recovery section. Three cut-out sections provided in the sides of the cover were designed to constrain the parachute in its compartment against negative decelerations and also to require minimal force

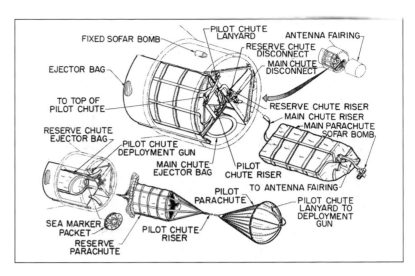

to break loose from its attachment at the time of deployment.

Pressure of the parachute pack would cause the cover to deflect so that attachment tabs pulled out from under attachment screw heads through a slotted hole provided for the purpose. The energy required to expel the drogue parachute from its compartment was provided from high-pressure gases generated by ignition

ABOVE A breakdown of the parachute systems and associated equipment installed in the recovery compartment. *(McDonnell)*

BELOW An excellent view of the top of the recovery compartment as Alan Shepard is winched aboard the recovery helicopter. *(NASA)*

BELOW A chart recording test results from numerous parachute drops and simulated abort conditions placing the canopies under various levels of stress. Logging different reefed conditions, the chart shows force versus velocity in drop tests carried out by Radioplane, the division of Northrop Corporation responsible for the parachutes. *(Radioplane)*

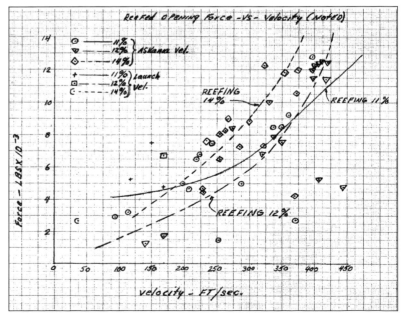

ABOVE Not all recoveries went according to plan. When the explosive hatch on spacecraft 11 accidently blew, the interior filled with water. Despite its best efforts the helicopter was unable to save it from sinking when a faulty warning light incorrectly indicated an overstressed engine and caused the helicopter crew to drop the capsule.
(NASA)

of a cartridge loaded with 66 grains of powder contained in a propellant can attached to a steel body which housed the ignition wiring and terminated in an electrical connector. The ignition circuit consisted of two separate and individual bridges, either of which was capable of igniting the powder on application of a current.

The main parachute consisted of a canopy, riser, deployment bag and parachute disconnect. The main canopy was a 63ft (19.2m) diameter ringsail type, essentially a slotted canopy similar to the ringslot parachute. It was fabricated from 2.25oz/yd^2 and 1oz/yd^2 nylon parachute cloth in 48 gores with 48 associated suspension lines of 550lb (249kg) tensile strength. It was packed in a deployment

bag that provided a low snatch force and orderly deployment. Manufactured from cotton sateen fabric, it was reinforced with nylon webbing and covered at the upper end with Thermoflex and glass cloth insulation. Inside the bag, midway along its length, were a pair of transverse locking flaps which would separate the canopy fabric from possible entanglement with the lines and facilitate full-time stretch-out before the canopy deployed.

Both the main and the reserve parachutes were attached to the spacecraft by a device designed to sustain parachute loads during descent and to disconnect the parachute at splashdown, a function necessary to prevent the capsule being dragged along the surface of the water after landing. The parachute riser was looped around the arm that transmitted the load to the structure through the piston. The shear pin restrained the piston from any motion that threatened to displace it. On impact, the electrical impulse from an inertia switch reached the squib cartridge causing it to fire. The gas pressure resulting from this forced the piston forward into the arm recess, cutting the shear pin in the process. Full displacement of the piston removed the transmission of parachute loads to the structure and allowed the arm to rotate around the pivot pin. The loop of the parachute riser then slipped off the arm and the disconnect function was completed. The lead buffer in this assembly

RIGHT Recovery was not a risk-free procedure. After surviving a ballistic shot to space and back, Gus Grissom narrowly escaped drowning when his suit filled with water – his flailing arms were at first taken to be a sign that he was safe!
(NASA)

served to absorb energy from the moving piston and prevented a rebound of the piston back into the locked position.

The reserve parachute consisted of deployment gun and lanyard, pilot parachute, reserve canopy, deployment bag and reserve disconnect assembly. The reserve parachute deployment bag was similar to the main parachute bag but with the addition of flaps at the upper end to contain the packed pilot parachute. The reserve parachute was identical to the main parachute canopy.

The pilot parachute was a flat, circular type, 72in (183cm) in diameter and with a 30ft (9.14m) bridle. It was manufactured out of 3.5oz/yd^2 fabric in the canopy and 2.25oz/yd^2 in the vanes. Initiation of deployment began with firing of the deployment gun by either electrical impulse or gas pressure, expelling a 12oz projectile that was attached to the reserve parachute pilot parachute. The pilot canopy would inflate and in turn pull out the reserve parachute. A one-second time delay was provided between receipt of the impulse and detonation of the main charge. This delay allowed the main parachute, if deployed or damaged, to separate from the capsule to avoid entanglement with the reserve parachute about to be deployed.

The gun was basically a tubular body that contained the main firing cartridge and the projectile assembly, the latter held in place by

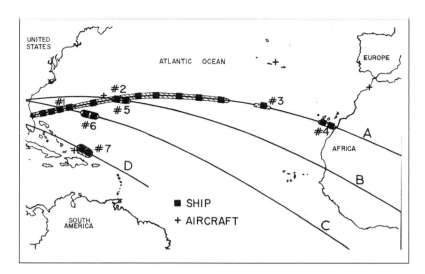

ABOVE Ground tracks of the first several orbits after launch indicate the assigned location of optional recovery zones. Continuous shaded area represents initial injection period, with A, B, and C indicating the first, second and third orbits. D is the planned recovery area after three orbits. In Area 1 the maximum access time for recovery forces to reach the capsule is three hours; in Areas 2, 3 and 4, recovery time is six hours. (NASA)

a pin that was sheared when the projectile was expelled. The main cartridge, which generated the gas pressure to eject the projectile, was fired through a firing mechanism that drove a firing pin into the primer cap at the base of the main cartridge, initiating a time delay train and causing the detonation of the charge. A minimum of 750lb/in^2 (5,149kPa) was required

LEFT The recovery equipment improved with each mission, and procedures and protocols changed in response to preceding experience. Here, Gordon Cooper remains aboard his pitching spacecraft while a flotation collar damps the worst effects of the swell and an open boat stands by. (US Navy)

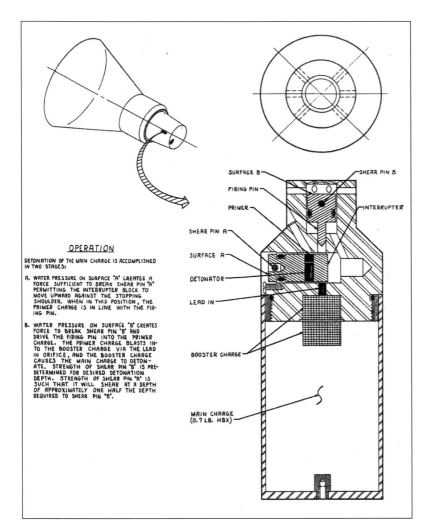

OPERATION

DETONATION OF THE MAIN CHARGE IS ACCOMPLISHED IN TWO STAGES:

A. WATER PRESSURE ON SURFACE "A" CREATES A FORCE SUFFICIENT TO BREAK SHEAR PIN "A" PERMITTING THE INTERRUPTER BLOCK TO MOVE UPWARD AGAINST THE STOPPING SHOULDER. WHEN IN THIS POSITION, THE PRIMER CHARGE IS IN LINE WITH THE FIRING PIN.

B. WATER PRESSURE ON SURFACE "B" CREATES FORCE TO BREAK SHEAR PIN "B" AND DRIVE THE FIRING PIN INTO THE PRIMER CHARGE. THE PRIMER CHARGE BLASTS INTO THE BOOSTER CHARGE VIA THE LEAD IN ORIFICE, AND THE BOOSTER CHARGE CAUSES THE MAIN CHARGE TO DETONATE. STRENGTH OF SHEAR PIN "B" IS PREDETERMINED FOR DESIRED DETONATION DEPTH. STRENGTH OF SHEAR PIN "A" IS SUCH THAT IT WILL SHEAR AT A DEPTH OF APPROXIMATELY ONE HALF THE DEPTH REQUIRED TO SHEAR PIN "B".

SURFACE B — SHEAR PIN B
FIRING PIN
PRIMER — INTERRUPTER
SHEAR PIN A
SURFACE A
DETONATOR
LEAD IN
BOOSTER CHARGE
MAIN CHARGE (0.7 LB. HBX)

ABOVE The SOFAR bomb, designed to detonate at a pre-set depth to send a sonic wave that could be picked up by a distant ship to plot the location of a spacecraft far from its predicted landing zone. *(McDonnell)*

for pneumatic operation. An alternative firing mechanism would have been to send an electrical impulse to the time delay igniter installed through the side of the gun. After a one-second delay the igniter would fire through the wall of the main cartridge and instantly detonate it. The ignition circuit consisted of two individual bridges terminating in a four-pin receptacle. The muzzle velocity of the projectile was 250–300ft/sec (76–91m/sec).

The parachute ejector bags were essentially inflatable air cells fabricated from lightweight rubberised nylon fabric, assuming the shape of a cylinder 11in (28cm) in diameter and about 35in (89cm) long. The upper end of the bag was slanted at full inflation to encourage jettisoning of the parachute pack overboard on landing impact. The parachute gas generator was a device to ensure a rapid and sufficient volume of gas to inflate both the main and reserve parachute ejector bags.

The reserve parachute gas generator was similar to that used for the main parachute with the exception of a one-second delay in ignition time. The generator functioned to provide gas through the relatively slow burning of a solid powder propellant in the main chamber. The gas was directed from the main chamber into the ejector bags through a 0.37in (0.94cm) diameter stainless steel tube, which also served as a heat exchanger to reduce the temperature to within tolerable values prior to entry into the ejector bag. The body was equipped with lugs for mounting the parachute container with four bolts.

Aids

As a post-splashdown aid to visual identification of the spacecraft a dye marker packet was incorporated which functioned by dissolving in water to produce a visible yellow-green patch. It consisted of 1lb (0.45kg) of fluorescein dye packaged into a soluble plastic bag inside an aluminium container. The dye formed a spot on the surface that would be visible from an altitude of 10,000ft (3,048m) and from a distance of 10 miles (16km) on a clear day. One preflight handling precaution to technicians was that it was to be kept in the dry and never exposed to water.

An additional aid to visually locating the capsule after landing was a flashing light installed in the recovery compartment. The intensity of the light was such that it would be visible in darkness for 46 miles (74km) and to an altitude of 12,000ft (3,600m). Powered by self-contained dry cell batteries, the light was energised by contact with the post-landing system relay activated by the closing of the inertia switch on impact, energising the impact relay. Flashing at a rate of once every 15 seconds, the light would operate for about 28 hours.

Another post-landing recovery aid was deployed at ejection of the main parachute after splashdown. Known as SOFAR (sound fixing and ranging) it detonated underwater when hydrostatic pressure at a pre-set depth ignited a small 0.7lb (0.3kg) charge of High Blast Explosive (HBX) that sent a shock wave through the water. This charge could be set to detonate at any predetermined depth as it sank

but usually at 3,500ft (1,067m), which would produce sound waves detectable to a range of 3,000 miles (4,800km). Sound detection devices aboard picket ships and shore bases would obtain a fix on the point of explosion and help recovery forces vector on the spacecraft.

For high-frequency voice communication a receiver-transmitter, HF recovery beacon and a whip antenna were provided. The active element comprised the collapsed antenna stowed in the recovery compartment that would be extended to 16ft (4.9m) after splashdown. It would be deployed on the action of a gas cartridge activated by the post-landing system power-drop 30-second time delay relay energising a whip antenna extension relay. While it was extending, a galling action between separate segments of the whip antenna held it rigid in the extended position.

Activation of descent and landing sequences utilised two pairs of baro-switches, located in the recovery compartment, and an inertia switch. The baro-switch worked with an over-centre spring to minimise chatter during vibration and shock and to prevent contact oscillations. One pair was set to close at 21,000ft (6,400m), initiating drogue parachute deployment, the other pair at 10,600ft (3,230m) ±750ft (229m), triggering fairing separation and main parachute deployment. The inertia switch was a spring-loaded device set to react to a landing shock of 7.5g, ±1.13g, to produce momentary closure of two electrical contacts to complete an energised circuit. The switch was used in conjunction with a latching relay that received an electrical pulse which, by latching into a locked position, provided a continuous supply of electrical power. The switch consisted of four separate snap-action switches and two separate reactive masses, together housed in one case.

While the normal operating profile of a Mercury mission had the capsule landing on water close to a full complement of recovery ships and capsule retrieval aids, provision was made for an emergency landing in a remote area where the astronaut would have to survive for prolonged periods. Electronic and visual aids were automatically activated at splashdown to aid in locating the capsule. The astronaut then had the option either of remaining inside his spacecraft or egressing to a life raft provided.

The supplementary survival kit was usually mounted to the back of the main instrument panel and contained an inflatable life raft stowed in collapsed condition, a water container with mouth-operated demand valve, a knife, a first aid kit, matches, a desalination kit, 10ft (3m) of nylon line, a whistle, a soap bar, a Search And Rescue And Homing (SARAH) beacon, a dye marker, a signal light, zinc oxide cream, a signal mirror, shark repellent, sunglasses held in a foam-lined case and a battery.

Some changes were made according to mission objectives, one example being that

ABOVE Lifted to an outer elevator deck of the aircraft carrier USS *Kearsarge*, Gordon Cooper rides spacecraft 20 before extricating himself through the side hatch. *(US Navy)*

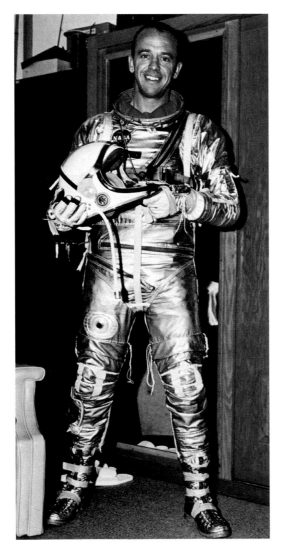

for the MA-3 flight launched on 25 April 1961, where the reserve parachute was replaced with a flotation pack attached to the main parachute in such a manner as to act as a buoy after main parachute disconnect to enable recovery of that canopy. Three dye markers were installed in the antenna fairing, which also remained attached to the main parachute. The markers were used to help locate the floating parachute and fairing.

Should an abort be necessary during the ascent to orbit but after the jettisoning of the escape tower the optional activation methods were similar to those required when aborting after lift-off, despite the consequences being somewhat different: ground command receiver abort signal; astronaut-activated abort handle; and booster catastrophic failure detection system. Any one of those would activate the Mayday relays. From whichever source, the relay would also be transmitted to the offending launch vehicle to shut down the sustainer engine. Through the contacts of the energised relays, a power circuit would be completed to the 'ABORT' light on the main instrument panel and the 0.20g contacts of the thrust cut-off sensor would be armed.

As the thrust decayed the contacts would close and power would be applied to the capsule separation delay relay after a one-second pause. After this the capsule separation bolts relay together with the capsule ring interlock relays would be energised and five capsule separation bolt squibs would fire, separating the capsule adapter clamp ring. The sequence from this point was the same as that involving the launch escape tower except that the escape rocket would no longer be in the loop. If the capsule had not reached orbital velocity the quickest way to abort would be to fire the retro-rockets. If that was not deemed advisable, separation of the retro-package would have to be undertaken manually.

Astronaut systems

Because its design began before that of Russia's Vostok capsule, Mercury was the world's first spacecraft designed to carry astronauts into orbit. The requirements were unique and without precedent and there was nothing from which engineers could receive guidance.

RIGHT A view looking aft showing the location of the astronaut in his couch inside the pressurised cabin. Relevant equipment, subsystems and systems are identified as to their location adjacent to his position. *(McDonnell)*

CENTRE A view looking forward from the astronaut's position, past the instrument console and forward to the escape hatch. Note the fuse panel to the left. *(McDonnell)*

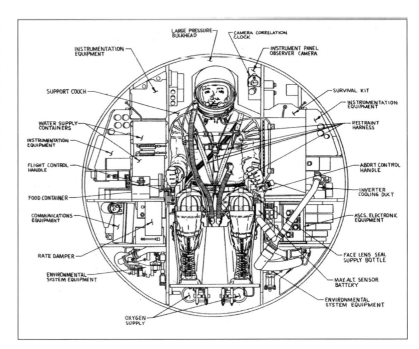

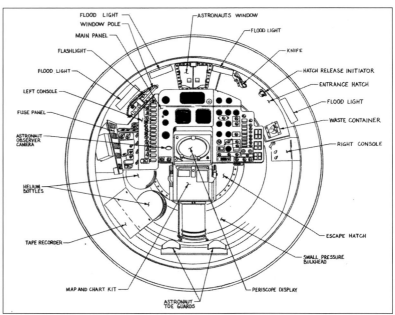

Cockpit and couch

One of the biggest challenges facing Mercury spacecraft design engineers was building a vehicle which was sufficiently flexible for a human to integrate with autonomous systems and to work with them intelligently to circumvent errors and failures, malfunctions which could, barring human intervention, result in the loss of life. Moreover, this argument fed back into the basic principle of engineering: KISS ('Keep it simple, stupid!'). In other words, use only essential systems and subsystems for mission success and the safety of the pilot; disengage from cluttering the operational running of the spacecraft through over-engineering redundancy in the never-ending search for enhanced reliability. There is a fine line between providing several layers of redundancy to improve reliability and making the system over-complex and prone to malfunction, eradicating all the best intentions of a 'safety-first' approach.

The interior layout of the pressurised cabin was designed so that the instruments, displays and controls were accessible to the astronaut and functionally integrated for minimum interference with mission tasks. With a pressurised volume of 55ft³ (1.55m³) the available space for the astronaut was about that in the cockpit of a small combat fighter. Only minimal attention was paid to ergonomic considerations, the objective being to provide a safe and secure environment for a single pilot to support the mission goals in as effective a manner as possible.

Essentially, the cabin contained a support couch, a restraint system, instrument and display systems, navigation aids, flight and

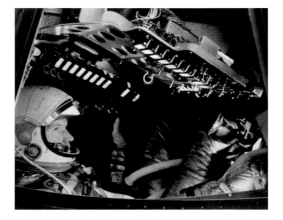

LEFT Gordon Cooper gives scale to the cramped confines of the spacecraft interior, affording about as much room as a fighter jet aircraft of the period. Note the bulky push-switches on the right-hand panel and the long flick-switches. *(NASA)*

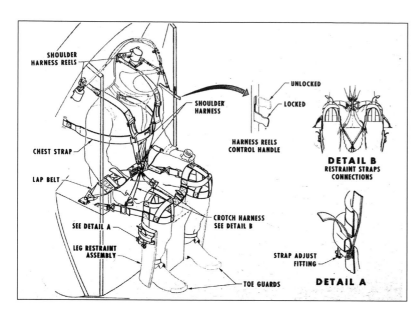

ABOVE At first the harness restraint system for the Mercury astronaut was a complex design that included leg guides and harness encapsulating the knees. *(McDonnell)*

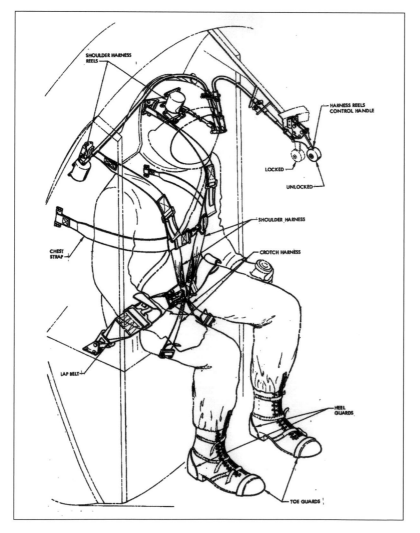

abort control handles, food and water supply, waste container, survival kit, cameras and electronic equipment necessary for operating the communication system.

The support couch – a form-fitting contour support tailored specifically for the precise body shape of each astronaut – was used from launch to landing. It was also there in the eventuality of the astronaut losing consciousness during peak accelerations and during abort or re-entry; only minor fluctuations in entry attitude, or in the gamma (flight path angle) of the re-entry trajectory, could considerably increase the physical loads experienced by the astronaut. The couch was located centrally within the large pressure bulkhead and consisted of a crushable honeycomb material bonded to a fibreglass shell and lined with a thin, but soft, comfort layer. The couch itself was fabricated in separate sections that had to be taken in through the hatch and assembled inside the capsule.

Design of the definitive form of couch took place in association with tests conducted on centrifuges where volunteers were subjected to high accelerations to duplicate the conditions an astronaut would experience at various phases of the mission. On one test, Lieutenant Carter C. Collins withstood a peak force of 20.7g for six seconds applied transversely, but astronauts would not usually experience more than 12g, and that only during re-entry following a ride on the Redstone rocket during a suborbital ballistic shot. The Redstone itself would impose an acceleration no greater than 6g during ascent. Usually a re-entry gamma of 1.5° would present acceleration forces of 9g but even modest changes in the entry angle would cause the g-loads to climb inexorably.

Some work on accelerations and g-forces on the human body had been carried out at Air Research and Development Command, and during April and May 1959 McDonnell carried out research to study the effects of high g-loads on living animals, with a view to discovering the

LEFT The standard harness changed in several subtle ways over the course of the Mercury programme but this arrangement was typical and was specifically applicable to the MA-9 spacecraft. *(McDonnell)*

degree of internal damage to vital organs which might occur. Four Yorkshire pigs were strapped into landing couches for tests of the crushable honeycomb structure and experienced 38–58g in drop tests from which they got up and sauntered off. It was reported by wags at the STG that the Mercury astronauts were not totally reassured by these tests!

The astronaut restraint system for the Mercury cabin was designed to firmly grip the body and prevent any accelerations in any axis from any loads the flight might incur. It consisted of shoulder and chest straps, a crotch strap, a lap belt and toe guards. The lap belt and crotch strap supported the lower torso, the chest and shoulder straps restrained the upper torso and the toe guards restrained the feet; the astronaut's hands and arms were restricted by gripping the abort and flight control handles located adjacent to the ends of the support couch armrests.

The instruments and displays for the attention of the astronaut were situated on the main instrument panel supported by left and right consoles. The main console was positioned so as to be directly in front of the pilot's support couch and was attached to the periscope housing. The main panel was designed so that the periscope display formed the lower control section of the instrument displays. The navigation instruments were placed on the left and centre sections while environmental system indicators and controls were located in the upper right section of the main panel. The electrical switches, indicators and controls for the communication system were situated on the left side of the main panel.

The left console was arranged so as to provide accessibility and visibility to the astronaut when in the fully restrained position. The console incorporated a telelight sequence and warning panel, the indicators of which are referred to throughout the systems explanations in these chapters. It also carried indicators and controls for the Automatic Stabilization Control System (ASCS), for the Environmental Control System and for the landing and recovery system.

The right-hand console, positioned below the side entrance hatch, also included controls for the ECS. A window pole device was located adjacent to the left of the observation window

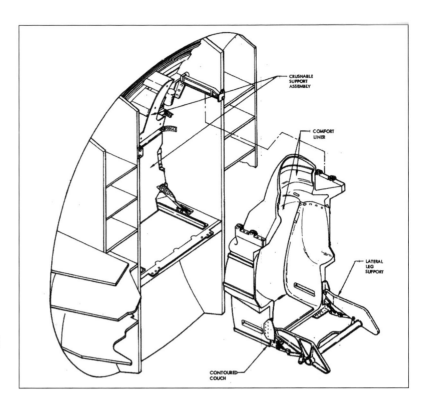

ABOVE The couch incorporated a thin comfort layer and was installed between the vertical structural frames attached to the pressure bulkhead. Recess shelves shown here were for batteries and other spacecraft systems. *(McDonnell)*

BELOW Joe Algranti evaluates the general configuration of the couch. This one is a test and evaluation mock-up and does not have the personalised, body-tailored, form-fitting shape astronauts used. John Glenn played a major role in his cockpit layout assignment, helping engineers design an ergonomic layout for access under the restraint of a fully pressurised suit in the event of an emergency. *(NASA)*

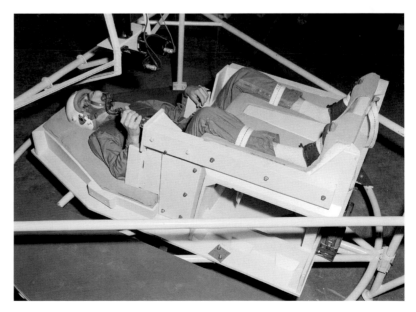

RIGHT Alan Shepard displays the suit harness and strap attachment buckles as he approaches spacecraft 7 for the MR-3 flight. *(NASA)*

BELOW Yorkshire pigs were selected for the job of testing living animals under very high g-loads, with a special couch and restraint system holding them in the supine position. This proposed restraint system was intended for Little Joe flights that never happened. *(NASA)*

that allowed the astronaut to operate controls while in a full pressure suit. The cabin was equipped with additional navigation aids and instruments to allow the astronaut to compute various control factors relevant to the flight or to landing. The navigation aids mostly consisted of the periscope, the satellite clock, Earth path indicator, altimeter, angular rate indicator and map case. All the navigational aids were situated in locations directly in front of the astronaut, either on or alongside the main display console.

Spacecraft controls were located at the forward end of each armrest on the support couch. The emergency support handle was situated forward of the couch left armrest and was used to initiate the abort sequence. As described elsewhere, to save accidental activation of the abort sequence the handle incorporated a manual lock. The manual control handle attached to the end of the right armrest was used for direct control through the Rate Stabilization Control System (RSCS) and its use and operation is fully explained in the 'Attitude control systems' section starting on page 88.

Periscope

Although the mission sequence of a Mercury flight was programmed into the spacecraft prior to launch, the astronaut was provided with a suite of essential tools and techniques, using the same systems, to compute altitude, course, velocity and landing information as well as to maintain the correct attitude of the vehicle in orbit.

Central to a lot of derivations by the astronaut, the periscope was installed to provide a clear view of the Earth below and to provide important navigation information. It was also used as a crude measurement of the spacecraft's altitude. It was designed by Perkin-Elmer, weighed 43lb (19.5kg) and included a wide-angle lens that extended or retracted through the external surface of the spacecraft. It presented the view to an 8in (20.3cm) screen mounted inside the pressurised compartment about 2ft (0.6m) from the astronaut's head. Altitude was determined by manually adjusting a knob that moved four indices on the screen until the apparent diameter of the Earth circle was framed. A calibrated dial read out the altitude directly.

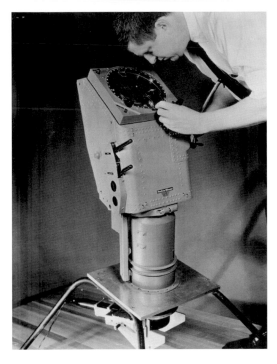

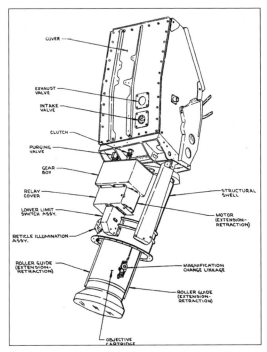

FAR LEFT The periscope system was a sophisticated piece of engineering that borrowed technology from the latest optical developments with astronomical telescopes, and pioneered new standards in precision alignment and mirror design. *(David Baker)*

LEFT The optics and the objective cartridge were held within a rigid structural shell that was fixed to the spacecraft and tested against thermal distortion or aberration. Its purpose was to provide a telescopic view of the ground below for navigational purposes, which in the case of Mercury meant that measurements and readings could provide the astronaut with basic information for calculating and operating a full re-entry sequence should all communications with the ground be lost. *(McDonnell)*

LEFT Internally, the periscope consisted of a series of mirrors, lenses and filters with separate adjustable-power lenses capable of giving the astronaut different magnifications. *(McDonnell)*

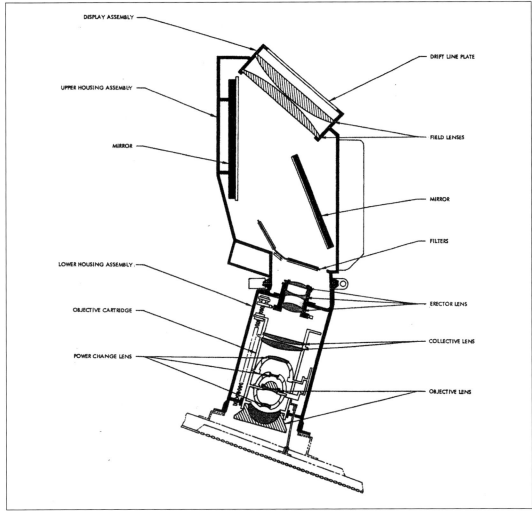

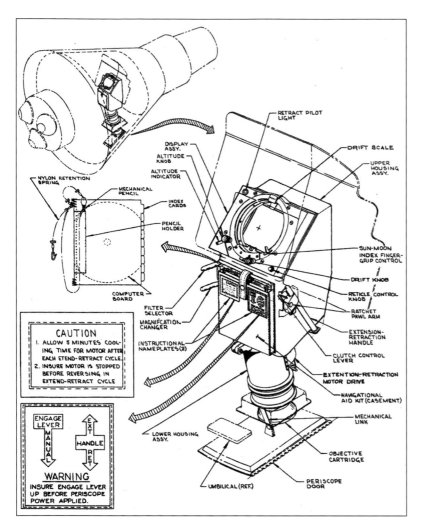

The schematic includes the following labels:

RETRACT PILOT LIGHT

DRIFT SCALE

UPPER HOUSING ASSY.

DISPLAY ASSY.
ALTITUDE KNOB
ALTITUDE INDICATOR

NYLON RETENTION SPRING

MECHANICAL PENCIL

INDEX CARDS

PENCIL HOLDER

SUN-MOON INDEX FINGER-GRIP CONTROL

DRIFT KNOB

RETICLE CONTROL KNOB

RATCHET PAWL ARM

COMPUTER BOARD

FILTER SELECTOR

MAGNIFICATION CHANGER

INSTRUCTIONAL NAMEPLATES (2)

EXTENSION-RETRACTION HANDLE

CLUTCH CONTROL LEVER

EXTENTION-RETRACTION MOTOR DRIVE

NAVIGATIONAL AID KIT (CASEMENT)

MECHANICAL LINK

LOWER HOUSING ASSY.

OBJECTIVE CARTRIDGE

PERISCOPE DOOR

UMBILICAL (REF.)

CAUTION
1. ALLOW 5 MINUTES COOLING TIME FOR MOTOR AFTER EACH ETEND-RETRACT CYCLE.
2. INSURE MOTOR IS STOPPED BEFORE REVERSING IN EXTEND-RETRACT CYCLE

ENGAGE LEVER
MANUAL
EXT
HANDLE
RET

WARNING
INSURE ENGAGE LEVER UP BEFORE PERISCOPE POWER APPLIED.

ABOVE This schematic shows how the periscope was mounted on a fixed part of the instrument display panel but with a manual retraction capability built in. The periscope was retracted and the door closed for ascent to orbit and for re-entry.
(McDonnell)

Several other indices on the screen provided navigational data such as the spacecraft attitude, its axial position relative to the Earth, and relative bearings of the Sun and the Moon. If the automatic re-entry control system did not function properly this information would help the astronaut to judge the proper moment for manual retrofire. The periscope was also fitted with a series of interchangeable filters to provide comfortable viewing under all ambient lighting conditions.

The periscope was designed to withstand shock stresses of up to 100g and consisted of three primary elements: the display assembly, the upper housing assembly and the lower housing assembly. The periscope was mounted in the spacecraft so that the objective cartridge could be extended or retracted through a door in the bottom of the spacecraft (relative to the seated position of the astronaut). The action of opening and closing the door was synchronised

with the extension or retraction of the cartridge by means of a mechanical linkage.

Preflight periscope preparations began on the launch pad when the instrument was extended; retraction was automatic with umbilical disconnect, the telescope only extending automatically again in the event of a 'hold' in the countdown from that point to ignition of the rocket motors. Under normal operations it would remain retracted until separation of the capsule in orbit and it remained so throughout the mission until 30 seconds after retro-package jettison, whereupon it was retracted again. At an altitude of 10,000ft (3,048m) the periscope was extended again and remained in this position up to the point of splashdown.

Relevant indicator lights to notify status included a 'RETRACT SCOPE' telelight on the left side console attached to the main instrument display. This switch had 'AUTO' (automatic) and 'MAN' (manual) positions but if the light did not go out within 40 seconds of retro-package separation it was necessary for the astronaut to place the switch in 'MAN' mode and retract the telescope that way.

As indicated earlier, the display assembly was required to provide information on the capsule's position relative to the Earth, providing details of drift, altitude, pitch, roll, true vertical, retro angle, field of view of the Earth-sky camera and relative bearings of the Sun and Moon. Mounted in the top of the upper housing, the display assembly incorporated scales, indicators, altitude reticles, attitude reticles and controls for their manipulation, Earth image, horizon image and the retraction pilot light. When the Earth was centred in the display the portion of the image bordering the altitude reticles would depict the Earth's horizon. The centre of the view, where the reticle lines crossed, represented a vertical line to the centre of the spherical Earth. The way the image was set up allowed the horizon and straight-down views to be true.

In the true vertical attitude, as compared to the optically vertical, the longitudinal axis of the capsule was perpendicular to a line passing through the spherical Earth. The capsule was optically vertical with respect to the Earth when it was pitched nose down at 14.5°. With the spacecraft in the optically vertical alignment the attitude of the capsule could be determined by means of a mechanism that included four pairs of altitude reticles which together formed a square within the circular display area. Turning an adjacent altitude knob shifted the reticles to change the size of the square, a function that simultaneously and proportionally rotated the scale of the altitude indicator.

To determine altitude the astronaut would rotate the knob, changing the shape of the square until the image of the Earth was inscribed within the inner square formed by the reticles. The altitude would then be read from the altitude indicator. The point on the Earth's surface vertically below the capsule would be indicated by the crossline reticle that appeared at the centre of the display window. This was illuminated by a lamp, the brightness of which could be controlled by the reticle knob. The altitude measurement scale allowed height determination between 57.5 miles (92.5km) and 288 miles (463km). It was calibrated to an accuracy of ±5.7 miles (9.2km) within a range of 115–161 miles (185–259km) and ±11.5 miles (18.5km) at altitudes below 115 miles (185km), and within ±57.5 miles (92.5km) at altitudes above 161 miles (259km).

This altitude mechanism could also be used to determine the pitch and roll with respect to the optical vertical attitude. The distance between the parallel curves which formed the side of the altitude reticle square was equivalent to a 5° angle at an altitude of 132 miles (212.4km). Moreover, if the spacecraft was off in pitch attitude it would be impossible to adjust the reticle square so that both the fore and aft reticle curves (relative to the 180° and 0° positions on the drift scale) were tangent to the image of the Earth. If the spacecraft was off in roll it would similarly be impossible to place the left and right reticle curves tangent to the Earth.

To determine the approximate pitch angle the reticle square was adjusted so that the horizon image lay across either the fore or aft inner reticle curve by about the same amount that the opposing horizon edge lay inside the opposite inner reticle curve. Whether the image lay across the right or the left was determined by the direction of pitch, which was itself an indicator of vector divergence. The degree of pitch was also determined by comparing the distance that one side of the

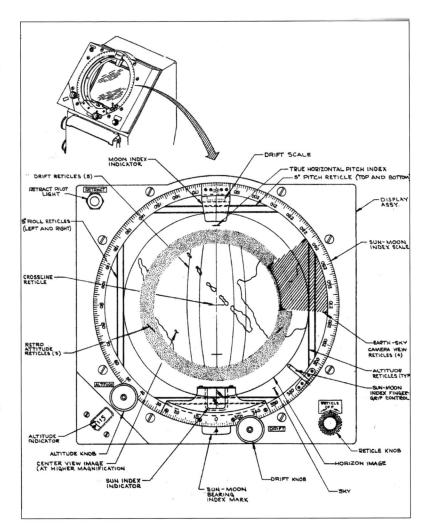

horizon image extended across the inner curve with the distance between the inner and outer curves. The approximate roll angle was also determined by the same method as that used for pitch angle.

Spacecraft drift rate was determined by reference to five parallel drift reticles etched on the face of the drift plate covering the circular display area. The drift plate itself would be rotated by means of the drift knob with a clear plastic scale mounted below the plate. The scale covered ±5° of drift in increments of one degree. It was necessary to set the periscope's magnification to high power to determine drift and the astronaut would rotate the knob until the ground track of the spacecraft, as viewed through the central high-power view, appeared parallel to the drift lines. The drift would then be indicated by the position of the centre drift line with respect to the scale.

To determine whether or not the capsule was

ABOVE The astronaut display screen was supported by indicators, reticles and drift scale, with Sun, Moon and altitude indicators for determining, to a reasonable level of accuracy, the altitude of the spacecraft above the Earth. *(McDonnell)*

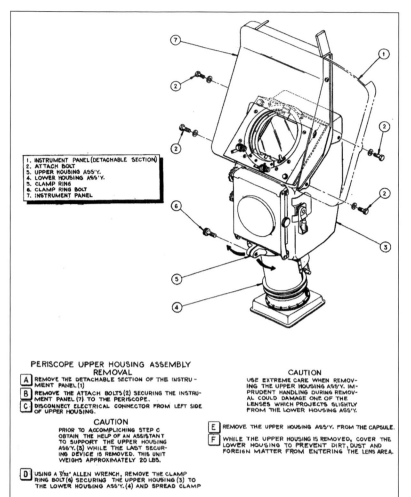

1. INSTRUMENT PANEL (DETACHABLE SECTION)
2. ATTACH BOLT
3. UPPER HOUSING ASS'Y.
4. LOWER HOUSING ASS'Y.
5. CLAMP RING
6. CLAMP RING BOLT
7. INSTRUMENT PANEL

PERISCOPE UPPER HOUSING ASSEMBLY REMOVAL

A REMOVE THE DETACHABLE SECTION OF THE INSTRU-MENT PANEL (1)

B REMOVE THE ATTACH BOLTS (2) SECURING THE INSTRU-MENT PANEL (7) TO THE PERISCOPE.

C DISCONNECT ELECTRICAL CONNECTOR FROM LEFT SIDE OF UPPER HOUSING.

CAUTION

PRIOR TO ACCOMPLISHING STEP C OBTAIN THE HELP OF AN ASSISTANT TO SUPPORT THE UPPER HOUSING ASS'Y. (3) WHILE THE LAST SECUR-ING DEVICE IS REMOVED. THIS UNIT WEIGHS APPROXIMATELY 20 LBS.

D USING A 7/32" ALLEN WRENCH, REMOVE THE CLAMP RING BOLT (6) SECURING THE UPPER HOUSING (3) TO THE LOWER HOUSING ASS'Y. (4) AND SPREAD CLAMP

CAUTION

USE EXTREME CARE WHEN REMOV-ING THE UPPER HOUSING ASS'Y. IM-PRUDENT HANDLING DURING REMOV-AL COULD DAMAGE ONE OF THE LENSES WHICH PROJECTS SLIGHTLY FROM THE LOWER HOUSING ASS'Y.

E REMOVE THE UPPER HOUSING ASS'Y. FROM THE CAPSULE.

F WHILE THE UPPER HOUSING IS REMOVED, COVER THE LOWER HOUSING TO PREVENT DIRT, DUST AND FOREIGN MATTER FROM ENTERING THE LENS AREA.

LEFT A page from the maintenance manual provides instruction on procedures to disassemble the periscope equipment. *(McDonnell)*

in the correct attitude for retrofire the astronaut would use the retrograde reticles. With the capsule set at the optically vertical altitude of 132 miles (212km), the nose of the capsule would be pitched down until the horizon image at the bottom of the display was tangent to the three retrograde reticles that formed an arc across the lower half of the display area.

The astronaut was provided with a means by which he could determine the relative bearings to the Sun and the Moon through a dedicated ring-shaped index scale mounted to the frame of the circular display area. The ring would be rotated manually by means of a finger-grip control. The inner scale was calibrated from 0° to 360°, with the Sun indicator mounted to the 0° position and the Moon indicator at the 180° position. Bearings would be read off the Sun-Moon index scale at the bearing index mark located on the main display assembly directly in front of the astronaut.

The astronaut was able to compute orbital velocity with the aid of a crossline reticle. The astronaut would start a stopwatch as

RIGHT The optical cartridge of the periscope assembly provided a view pitched 14.5° forward of the true vertical. This allowed the astronaut to use it for angular alignment checks, both on the spacecraft attitude readouts and for deriving this value in the event of an instrumentation failure. *(McDonnell)*

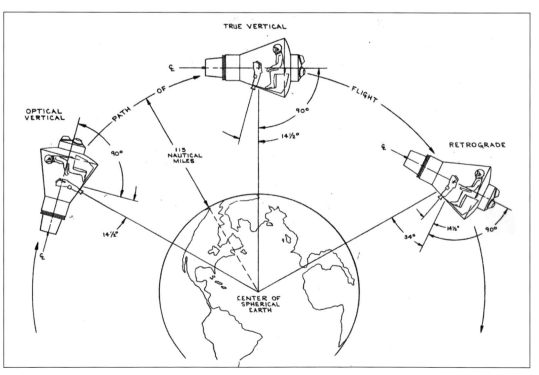

tho first designated object passed under the crossline reticle and would stop it as the second checkpoint passed under the small crossline reference. With the elapsed time the orbit could be calculated using the spacecraft hand computer.

Included in the navigational equipment was an Earth-sky camera. Four camera reticles were located at the right centre of the display area to provide an approximate outline of the camera image that was included in the spacecraft equipment.

The upper housing assembly of the periscope incorporated two mirrors, a filter installation, a magnification change control mechanism, a manual extension-retraction control, a housing exhaust valve and a desiccator, which included the housing intake valve and a housing purge valve. Within that structure the periscope mirrors were aligned with great precision and set at an optically convenient angle for use by the restrained astronaut, who had freedom of movement only with the direction of his head. The upper assembly also contained clear, red, yellow and medium neutral density filters mounted in a rack that could be rotated to position the selected filter in the optical path.

The periscope's manual extension-retraction mechanism enabled the objective cartridge to be driven in or out. It consisted of a manually driven gear system that was coupled to the gearbox and motor drive assembly in the lower assembly housing through a manually operated clutch. The system was driven by a ratchet handle situated on the right side of the upper housing, pulled up to place it in operation and pulled down for stowing. For manual manipulation the engage lever would be in the down position and a ratchet pawl would determine the direction of motion.

The intake and exhaust valve allowed the upper assembly to receive the oxygen atmosphere from the pressurised cabin without the passage of dirt or dust particles to the inside of the housing. The intake valve was attached to a structural housing to which the desiccant tube was attached. When cabin pressure exceeded that of the upper housing the intake valve would admit air through the silica gel. Moisture and dirt would be removed to reduce the level of condensation and prevent

deposits settling on the optical components. A separate purge valve provided an inlet and an attachment point for purging the upper housing with dry nitrogen gas.

The periscope's lower housing assembly consisted of a structural shell that contained the objective cartridge, with extension-retraction motor, gearbox, lower limit switch assembly and reticle illumination assembly mounted to the structural wall. The objective cartridge contained the lens, power change lens and the collective lens and was mounted inside the structural wall of the assembly housing. A connecting link between the objective cartridge and the periscope door allowed the latter to move in unison with the telescopic motion of the objective cartridge. The wide-angle objective lens, approximately 180°, gathered light that passed through to the collective lens where the first image was formed. The power change lens could be moved in and out of the optical path using a mechanical linkage, actuated by the magnification changer. For low power operation the power change would be moved out of the optical path but for high-power operation the lens would be moved directly into the optical path.

To support mission requirements and to assist the astronaut with timely support of a preconceived flight plan several supplementary

BELOW This shot of Freedom 7 on the sea clearly shows the periscope extended through the access door opened after re-entry. *(NASA)*

materials were necessary, which were kept in a navigation aid kit that consisted of a neoprene-coated nylon case attached to the periscope directly below the circular display area. The kit was essentially a binder with a variety of index cards. These were used to file checklists, rate cards and navigation charts that varied according to the requirements of particular missions. As flight plans were developed and became more formal, increasingly packed with engineering tasks and some scientific experiments, it was important that the operating sequences allowed the astronaut to take control of the spacecraft within the limitations of the engineering and its systems.

These were the developmental steps that began to transform the manner in which the pilot was addressed within the programme, and gradually, over time, it was realised that the astronaut was no longer a physiological specimen but a working test pilot for the new capabilities afforded by space flight. The manifestation of that transition was found in the contents of the navigation kit, which became the astronaut's link with everything that had been designed into the mission. Attached to the case were a pencil holder, a mechanical pencil and a nylon retention spring. The pencil holder was made from neoprene-coated nylon

and was sewn on to the case. The mechanical pencil was secured to the binder by a nylon retention spring and another spring of the same type secured the binder to the case.

Late in the Mercury programme an additional navigational aid was provided in the form of an optical device that determined whether the spacecraft was in the correct retrofire attitude by taking the essential provisions of the periscope and placing them within a device fixed in the rear and to the left of the main observation window. To save weight for the 22-orbit MA-9 mission, the 76lb (34.5kg) periscope was deleted and the new window-mounted device carried instead. The navigation reticle was mounted on a 180° axis from left to right so as not to obstruct the astronaut's field of view.

When rotated to the viewing position, it contained a tinted red light of sufficient brilliance to illuminate four red lines, three of which were vertical. The fourth line was horizontal, which was required to be tangent to the Earth prior to retrofire. The lighting system was automatically turned off when the instrument was in the stowed position but incoming light could be dimmed by a polaroid filter operated by rotating the outside of the instrument, which varied the intensity of the light.

Pressure suit

The design of the Mercury spacecraft inherited from the Air Force MISS programme established the requirement for a fully pressurised cockpit for the astronaut. But in the event of a depressurisation it would be necessary for the pilot to survive in a pressure suit, which in itself brought demands exceeding the abilities of any suit available in the late 1950s. The long and tortuous story of how research into high-speed and high-altitude flying stimulated the development of spacesuits has been told in great detail elsewhere but suffice to say that the requirements for Mercury were unique.

Cooperation between the then NACA, the US Air Force and the Navy began in June 1954 when representatives met to discuss the next generation of rocket research aircraft which were expected to fly to the very edge of space itself and achieve hypersonic velocities in vehicles preceeding boost gliders such as

RIGHT The pressure suit was a uniquely challenging part of the Mercury programme and resulted from bringing together several existing suit designs for high-altitude/high-speed aircraft and creating a concept suitable for space flight. Suiting-up took about an hour from start to finish. Here John Glenn dons the undergarment while a suit technician prepares the separate components. *(NASA)*

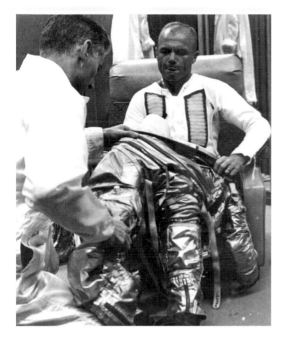

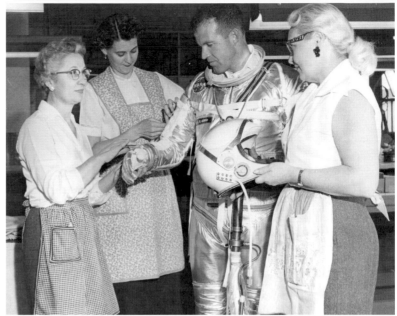

Dyna-Soar. Prior to that, the Air Force and the Navy had been cooperating since 1947 on developing high-pressure suits. The result was a suit for the X-15 created by the David Clark Company that evolved rapidly into the MC2 and then the A/PSS-2 which became standard for Air Force high-altitude tests.

The first Mercury suit conference was held on 29 January 1959 when more than 40 specialists gathered to discuss all aspects of the requirement. By the spring three companies had emerged as leads in detailed studies of a suit which would satisfy NASA's requirements for Mercury: David Clark Company; Playtex Division of the International Latex Corporation; and the BF Goodrich Company. Candidates from the three contenders were rigorously tested by respective participants and two finalists were chosen to go head-to-head, with the David Clark Company eliminated. NASA chose the Navy's Mk IV as the baseline from which to develop the Mercury suit.

The Mk IV employed a two-gas system with oxygen breathed in to the oral nasal cavity by an oxygen demand valve in the helmet. The helmet was separated from the larger suit cavity by a face seal inside the helmet. Exhaled gas was exhausted by a check valve into the main suit cavity, which was pressurised by compressed air from an aircraft cabin pressurisation system. The NASA Mercury suit would be a one-gas closed, recirculating

ABOVE LEFT Donning the one-piece leg and torso suit required careful adjustments to prevent disrupting the bio-harness worn for a flight. His undergarment displays the ventilation panels on the upper body part. *(NASA)*

ABOVE Gordon Cooper pauses for a posed publicity picture during a visit to the Goodrich Company where seamstresses pretend to make final adjustments to their handiwork. Women played a highly significant, if not always visible, role in many aspects of the Mercury programme, not least in the meticulous and tedious job of fabricating the hand-made pressure suits tailored to each astronaut. *(Goodrich)*

LEFT Gloves and boots went through some evolution during the programme, and their design changed as experience led to simplification and modification. This was another vital role for Mercury, as suit design advanced rapidly and led directly to the G3C suit for the Gemini programme. Note the circular mirror to reflect the image of the display panel to the camera observing the astronaut. *(NASA)*

pressurisation, ventilating and breathing system. Due to the finite quantity of oxygen on board great emphasis was placed on minimising suit leakage. Also, the pressure drop in the suit system had to be held to a minimum to avoid excessive cabin ECS system loss.

On 22 July 1959 Goodrich got the contract for 21 suits plus two spares. Four operational research suits were to be made for astronaut Walter M. Schirra, Dr Bill Douglas and twins Gilbert North and Warren North. Bill Douglas was the official Mercury flight surgeon while Gilbert North worked for McDonnell and Warren was at NASA headquarters in Washington DC. Nine suits were ordered for engineers and astronauts and a further eight were to be produced – one for pre-production qualification and seven flight suits for the astronauts. Considerable development, testing, re-testing and re-design went in to the evolution of the suit and not until the end of 1960 did the final qualification take place.

Immediately after receiving the contract in

BELOW John Glenn (left) chats with Gordon Cooper during suit-up preparation. Note the later style of glove, considerably more comfortable than the earlier design, which did not have the extra lacing adjustment that allowed personalised fitting according to the preferences of the astronaut. *(NASA)*

July 1959 to build the suit, Goodrich veteran suit designer Russell M. Colley set to work to modify the Navy's Mk IV design, which only the previous year had been selected by the Air Force for Arctic conditions. With a weight of 20lb (9kg) it was a one-piece garment with separate boots, gloves and helmet. Characterised by a diagonal, pressure-sealing zipper across the upper torso, it was fitted with an inner layer of Helenca stretch-knit fabric which gave it great flexibility and comfort when unpressurised. Additional work had already been done on pressure suits for high-speed and high-altitude aircraft when requirements for the U-2 and SR-71 produced novel and innovative adaptations. Several evolutionary lines of development converged from the programmes to create the Mercury space suit.

As designed, the Mercury suit had three components: the torso, the helmet and the gloves. Each astronaut was custom-fitted with a bespoke suit sized to the individual's unique dimensions, even down to the most personal piece of a man's anatomy usually approximated by the average tailor! The body section was double-layered with an inner gas bladder fabricated from neoprene and a neoprene-coated nylon fabric. The outer layer was a heat-reflecting aluminised nylon.

Various myths have surrounded the choice of silver as an exterior coating, Alan Shepard boasting that it was chosen by the astronauts themselves because it was associated with the derring-do of fictional space pilots in books and films. As usual, the truth is less poetic. Silver was chosen because of its thermal reflectance properties and was considered the optimum finish for the spacecraft, as it had been in the X-15 programme. With the introduction of the Gemini programme there would be no more silver suits.

'Wally' Schirra had been assigned to development support of the pressure suit and reported many shortcomings that were related to the suit's original Mk IV design and to an overly-conservative approach brought on through the X-15 suit, many aspects of which had been adapted for the Mercury suit with little success. During this period new materials, fasteners, zippers, connectors and fabrics were tried and tested. While the suit was not intended to be pressurised during a normal mission – it was there only as a backup – the difficulty lay in the compromise between an acceptable design and build-configuration for what was expected to be the normal unpressurised application. The requirement for it to support full mobility and limb flexibility when pressured in an emergency also compromised comfort when worn in the cockpit.

Compromise too was necessary to create a working design applicable to both normal and emergency applications, and throughout the 18 months of development and qualification work several modifications and revisions occurred. Mobility of limb movement was one area where the compromise extended into the design so that if full pressurisation took place there would be several uniquely placed break joints where cinched fabric at the elbows and knees would be released from a tethered condition. But the ballooning effect was so great that it made it almost impossible to flex arms and legs under a fully pressurised condition. The inner lining of the suit carried a series of tubes that carried oxygen down to the lower extremities and provided ventilation.

The helmet consisted of a hard resin-impregnated fibreglass shell for impact protection with a crushable impact liner, ventilation exhaust outlet, a visor sealing system and appropriate pick-off points for installation of a communication system. The helmet was held in position by straps attached to the neck and torso under full pressurisation. Except for the first manned flight, gloves were attached and locked to the lower forearm by a nylon-sealed ball bearing ring. Each glove had an inner and outer layer and middle and index finger-tip lights were provided powered by a battery on the back of the glove. The boots were of special lightweight nylon construction and like the gloves had lacings for comfort fit. Under the suit the astronaut wore a one-piece lightweight cotton garment and special patches were provided to improve the flow of air across the body.

During final suit acceptance procedures, several key tests were carried out which were designed to stress it for weaknesses, including dropping an 8lb (3.6kg) spherical steel weight on the apex of the helmet with an impact energy of 300lb (136kg) weight,

LEFT Particulars of the helmet changed over the relatively brief duration of the Mercury programme. Note the dual redundant helmet-mounted microphones and the improved helmet neck ring and clasp lock. *(NASA)*

BELOW Gordon Cooper brought the Mercury flight programme to a magnificent conclusion with the 22-orbit flight of spacecraft 20 (Faith 7) on the MA-9 mission. Here he displays the definitive Mercury suit improvements incorporating subtle modifications, improved helmet microphone system and better fittings. *(NASA)*

ABOVE Wearing protective overshoes, John Glenn is accompanied by Dr Bill Douglas (centre) and technician Joseph Schmidt. *(NASA)*

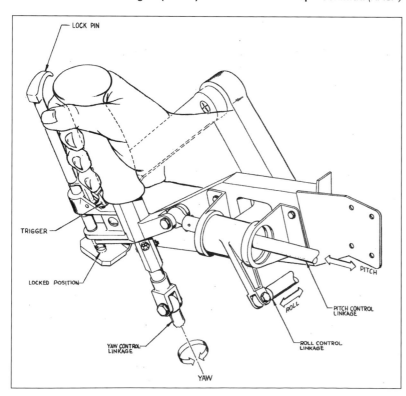

and subjecting the pressure garment to extremes of environmental exposure. Tension tests on hoses, visor seal cycle tests, multiple locking and unlocking cycles of the circular neck dam and helmet interface, leakage tests and extremes of environmental stress to the communication system were only a few of the rigorous stresses imposed on the pressure suit and its ancillary equipment.

The baseline suit configuration was only flown on the MR-3 flight involving Alan Shepard. Changes to the suit for Gus Grissom's suborbital MR-4 flight replaced the zipper closure disconnects with a self-donning bearing, adding a parabolic mirror to the chest and removing a slight 'peak' to the helmet. John Glenn's suit for MA-4 was the same except for smaller glove wrist buckles, removing the torso vent duct disconnect, installing the finger-tip lights and battery on both gloves, replacing Roanwell with Electrovoice microphones, adding special leak checks to the wrist bearing seal, and adding a life vest on the parabolic mirror.

Scott Carpenter's MA-7 suit replaced the moulded rubber gloves with Estane dripper bladders and added red colouring to the tip lights on the left glove. Wally Schirra's suit worn for MA-8 was the same but with an additional strain relief at the top of the pressure sealing closure, changes to the wrist disconnect lock material and replacement of the Electrovoice microphones with ones supplied by Plantronics. Gordon Cooper's MA-9 suit changed the Plantronics microphone to one from the same company incorporating noise-cancelling features, replaced the pneumatic visor seal with a mechanical seal and replaced the separately donned boots with ones integral to the suit legs.

Several changes resulted from near-disasters including the experience of Gus Grissom when he came close to drowning after his capsule

LEFT **The manual attitude control and stabilisation of the spacecraft was conducted by way of a rotational hand controller, installed to be accessible to the astronaut's right arm. Rotation of the hand controlled yaw motion, tilting the handle left or right adjusted roll, and pushing it in or out caused pitch deflection up or down respectively.** *(McDonnell)*

began to sink and he took to the water, only to find that his suit began to fill! The modest attempt to make an effective seal between the neck dam and the torso area had proved inadequate and changes were therefore made to ensure a more effective seal. In preflight simulations too, excessive bladder leakage was frequent and troublesome, with almost continuous repairs effected by the manufacturer. The location of the exhausted oxygen in the helmet caused eye irritation and a distracting noise and a mechanical closure was installed for MA-9 to prevent a visor seal leakage.

Many improvements were recommended at the end of the Mercury programme, both to improve comfort and mobility and to ensure a more reliable suit, with many recommendations carried along to the Gemini programme that followed Mercury into flight operations. But for all its work Goodrich was never to make another pressure suit flown in space, the David Clark Company getting the contract to produce the Gemini suit; but not before 1965 was NASA's definitive 'space' suit produced, the first to allow a US astronaut to leave his spacecraft in orbit and drift weightless free of its confines. Before that, the Russians had first demonstrated this leap in capability with the Berkut suit worn by Alexei Leonov on 18 March 1965 to float freely outside his Voskhod 2 spacecraft.

Biomedical sensors

The original idea of putting a man into space in a ballistic, recoverable capsule was driven by the need for physiological data, so after the requirement for a pressure suit had been established the next essential task was to design a system of sensors which could measure various parameters related to his health and physiological condition. One concern was the inability for anyone on the ground to react to an emergency in space. So it was vital that medical information should be transmitted to the ground when the spacecraft passed over a tracking station.

Countering the argument for an extensive suite of bio-physiological sensors was the need to keep weight as low as possible, so all the electrical components were miniaturised and this development programme alone would

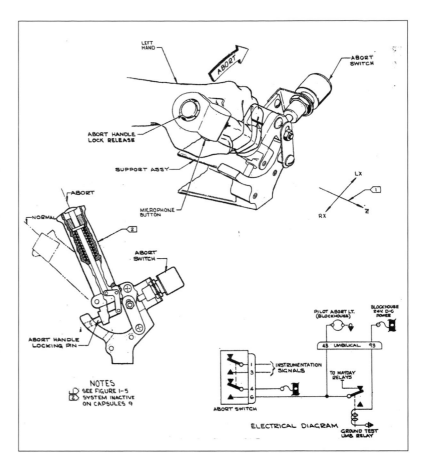

reap many benefits in the rewards gained for military aviation and the flying of high-speed combat aircraft. In Mercury three channels were allocated to biomedical indicators that provided data on body core temperature and respiratory rate and depth, and an electrocardiogram.

Selection was based on the common use of those three parameters in centrifugal simulations of the stresses of high acceleration and in altitude-chamber studies. With a set of 'ground truth' data, extrapolation into the weightless environment maintained an established set of protocols by which to compare the astronaut's responses preflight and in-flight. But preflight simulations were of short duration, and while the standard sensor transducers were adequate for a few hours at most they were insufficient for operation over the 24-hour period for which Mercury flights were envisaged.

Design requirements for new biomedical instrumentation were given to McDonnell and in support the Space Task Group at NASA developed several designs of their own for providing this information. Concurrently, an extensive search of the literature was made

ABOVE The manual abort demand was operated by the astronaut's left hand, the lever held securely by a lock release to enable him to push the handle forward for activation. *(McDonnell)*

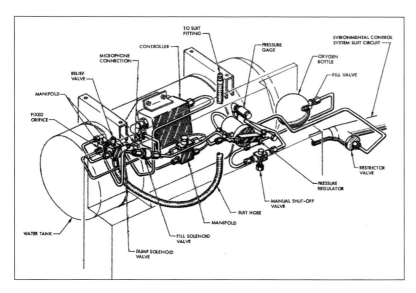

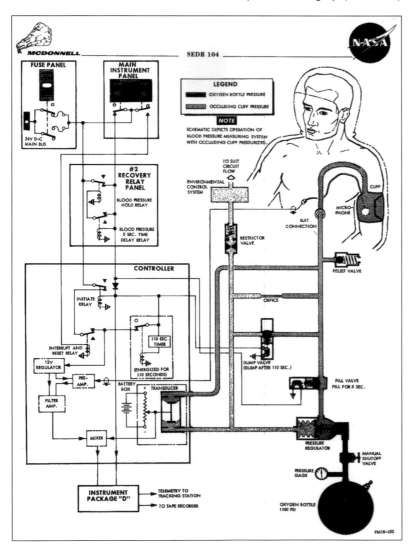

LEFT With physiological studies of the astronaut a primary objective of the entire programme, key parameters included measurement of respiration rate, heart rate and blood pressure, the last proving particularly difficult to achieve in flight. The blood pressure monitoring system eventually developed was integral to the medical evaluation of the pilot. *(McDonnell)*

and relevant details scavenged from a broad range of medical and scientific studies. New transducer designs challenged the technology of the day because nobody had designed equipment that was required to operate under such physical stress as the astronauts would endure during their missions.

Quite quickly in their training programme for space flight, the astronauts themselves were required to try various different sensor designs to see which worked the best and provided the most comfort. As training programmes subjected the Mercury astronauts to rigorous simulation and investigation of bodily responses, they themselves became the test subjects for new biomedical sensors. McDonnell investigated the performance of their own sensors in their dedicated altitude chamber but the STG ones were used at various other facilities involved in the training programme.

As the programme developed it was necessary to apply the new equipment to the primates which would precede man into the space environment, and the chimpanzees which were used for the flight-test programme also tried out the new sensors, so that physicians, engineers and technicians could evaluate the interface with spacecraft electrical systems and the quality of the data sent to the ground as telemetry.

Body temperature had long been an indicator of physiological condition in aviation and the optimum way that it had been measured was by the use of a skein of ten sensors situated around the body and bundled to a single outlet where a summed average temperature was used to derive the actual value. But this would be too limiting for the astronaut, with multiple leads from a range of sensory points restricting his movement. The solution was a rectal probe sensor utilising a thermistor. Most of the rectal probes in use at the time were inserted 4–6in (10–15cm) and were as much as 0.25in (0.6cm) in diameter, causing discomfort and psychological rejection. Detailed study revealed that even when inserted up to 8in (20cm) the probe end would tend to bend up or laterally and increase the discomfort while decreasing its value.

The search began for a suitable design where the probe would be inserted a more comfortable 2–3in (5–7.5cm) and another search got under way for a group of volunteers to try them out. The Langley facility came up with designs, as did McDonnell, and a range of commercial products were bought in. Test volunteers were required to wear them for up to eight hours, which considering the amount of time before and after the flight when they would have to be 'installed' was only the duration of a three-orbit mission. Several optional designs were tried out in the centrifuge and some details of rectal probes used in the X-15 rocket research aircraft programme helped improve the hardware.

As the tests increased the development cycle continued to come up with better designs and throughout the programme various changes were made. Boeing responded by developing a glass 'thermocup' about 2.5in (6.3cm) in diameter in the centre of which was a sponge-rubber mound that supported the thermister. The sensor was worn on the surface of the body on the axillary line, under the place where the arm joins the shoulder, and was held in place by a web belt. Until that was introduced, astronauts continued to suffer the indignity of a rectal probe.

The rectal probe was a 1,490-ohm thermister element that was utilised as one leg of a bridge circuit that formed the two inputs to two DC amplifiers, the output of which was applied to the telemetry transmitters. The 0V DC to 3V DC output represented a temperature range of 95°F (35°C) to 108°F (42.2°C), both amplifiers being calibrated R and Z by ground command.

Physicians were concerned about the astronaut's stress level through respiration and to measure this the standard Haldane method was preferable. A Scottish physiologist and philosopher, in 1905 John Scott Haldane discovered that the rate of breathing was regulated by the effect of the tension of carbon dioxide in the blood on the respiratory centre in

the brain. The measuring apparatus usually used for this analytical work involved exhaled gas collection, but this was bulky and heavy – quite unsuitable for use in the physical limitations of the Mercury spacecraft. The simplest method would be one that uses the pneumographic technique that employed a linear or rotary potentiometer but it was also possible to obtain data through a thermister sensor.

The original three candidate methods were tried. The potentiometer sensors were attached by a web belt around the chest, with the potentiometer worn at the nipple line, but these pneumographic sensors were unreliable because they could not prove that air was moving in and out of the body. The subject could deliver a false reading by tensing his muscles and could register a reading by chest contractions against a closed glottis. McDonnell designed a linear potentiometer but this impeded the subject's breathing and caused great discomfort. And then a novel solution was found.

By mounting the thermister on the lip microphone inside the helmet and heating it to 210°F (99°C) the movement of air was recorded to determine the respiration rate. It could not, however, give a reading of the volume of air moved by the exhaled process but it did provide the valuable information the physiologists sought. Several centrifuge runs were made with the device attached and it was found to work well, the flight unit containing a 6V excitation current to provide the temperature. Sensing recorded the dissipated heat across the mouth and the nostrils but if the astronaut moved his

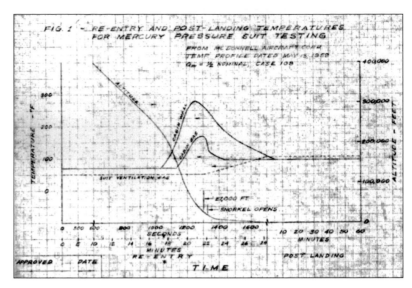

ABOVE Integrated with the biomedical monitoring of the astronaut's condition was a real need to understand the temperature environment of the spacecraft. This chart shows a very early, but consistently relevant, determination of the temperature profile of the pressure suit compared with the temperature of the wall of the spacecraft, the gaseous envelope of the pressurised compartment and the stipulated suit ventilation temperature.
(McDonnell)

head the ability to quantitatively measure the volume by derivation became impossible; the movement cycle was, however, still possible.

The measurement of the respiration rate applied an input power of 3V DC to the thermister attenuated proportionally to its changing resistance, which in turn was proportional to the magnitude of the astronaut's respiration rate. The rate itself was determined by the astronaut's breathing. The changing output voltage of the thermister varied the bias of a transistor and in-line amplifier, the output signal then being applied through a calibration card in the 'F' package of the instrumentation suite to two 1.3kC voltage-controlled oscillators in the 'D' package. The output of the VCOs was then applied to the High and Low frequency telemetry transmitters. The calibration card provided a means of interrupting the signal for R and Z calibration with sensitivity adjusted by a potentiometer in the electronic assembly.

Monitoring electrocardiograms from subjects in motion was gaining great popularity in the late 1950s and several types of ECG recording methods had evolved, but none were compatible with the Mercury spacecraft amplifier systems. It was, for instance, undesirable to use the standard electrode attached with a standard paste that carried an electrolyte that was hypertonic. The nature of this paste was that it would dry out and lose conductivity, the electrode-to-skin resistance increasing to cause considerable irritation. The electrode would also become less effective. Moreover, the Mercury ECG preamplifiers had an input impedance of 25,000 ohms whereas the human skin has an external resistance of 75,000–100,000 ohms. An added problem was finding a system that would operate effectively from subjects engaged in muscular movement.

The search began to find an electrolyte that would maintain a moist area without irritation for the full duration of a 36-hour mission and tests were eventually made on ten individuals with 36 different combinations for attachments and chemical compounds. Two 1in (2.54cm) diameter watch covers were filled with various electrolytes and placed on the skin, one attached to the back of the trapezium muscle and another to the chest lateral to the sternum. It quickly became obvious that key to their effectiveness was some sort of

seal to prevent it drying out. It was eventually decided that the best method was an electrolyte made from a modified bentonite compound. Bentonite is an absorbent aluminium phyllosilicate clay. The NASA compound consisted of 15gm of bentonite, 10cc of water and 3gm of calcium chloride.

The most commonly preferred form of electrode was the platinum suture which involved a fine wire stitched under the skin so that it pierced the epidermis with its high resistance and gained contact with the soft tissue beneath which had more conductive properties. But this invasive method was rejected because it required skin puncture, which was not acceptable on an astronaut. McDonnell was again responsible for the electrode development programme, though a parallel programme was set up at STG and several different types were evaluated at NASA's Langley Research Center. After many tests and simulations under space flight accelerations and thermal cycles the decision came down to one of three possible options: a stamped stainless-steel cup; a stainless-steel wire mesh; or a fluid electrode.

After further exhaustive tests the fluid electrode was selected for Mercury because it had more endurance, greater durability and required less attention during use. The Mercury ECG amplifier system was designed for three electrodes. However, if either the right side or upper chest electrode failed one system would still be transmitting, but if the left electrode came loose all three would be lost. A four-electrode system was introduced to prevent this and the precise locations for the electrodes were mapped after consultation with Dr James A. Roman and Dr Lawrence E. Lamb of the US Air Force School of Aviation Medicine and Captain Ashton Gabriel of the US Naval School of Aviation Medicine.

A new lead design was provided with an auxiliary line on the left and right axilla. The left electrode was placed at the base of the rib cage and the right electrode on the rib cage at the third intercostal space. The second lead, or sternum, was at right angles to the first with one electrode on the manubrium and the other on the xiphoid process. The new system was extensively tested on the centrifuge and in the simulators and proved highly effective and good results were obtained when manned flights began with Alan Shepard's flight in MR-3.

Signals from the ECG were obtained from four transducers, as described earlier, the outputs being applied to two amplifiers in D package with left and right side paired with one and upper and lower to the other. Signals were then directed to the 2.3kC and 1.7kC voltage-controlled oscillators which in turn applied their respective outputs to the High and Low telemetry transmitters. The 2.3kC VCO's input signals were divided between the ECG and blood pressure outputs.

When physiologists began planning for monitoring of the human body in Mercury flights they had hoped to be able to measure blood pressure and this was discussed extensively, but no effort was made to introduce it into the spacecraft until after the two suborbital flights in 1961. Specific work to develop a means of measuring blood pressure began around the time of the MR-3 flight in May 1961 and a system using a unidirectional microphone and cuff was introduced. Developed by AiResearch Manufacturing Division of the Garrett Corporation, it employed a 35-cycle filtering circuit inserted between the inflatable cuff and the subject's arm. The inflatable cuff was built into the pressure suit and incorporated a cylinder which cycled every 140 seconds. The pressure in the cuff system ranged between 250mmHg and 40mmHg.

The new system was tested on the centrifuge and a system worked out whereby it could be triggered by the astronaut or at fixed intervals with one of the ECG channels used intermittently for systolic and diastolic data. It was ready in time for the first manned orbital flight, MA-6, in February 1962. An automated version of the blood pressure monitoring system was introduced for the MA-7 flight but this showed suspiciously high values that triggered an exhaustive investigation as to the cause. It was determined that elevated rates were caused by instrumentation error and this prompted a specific plan for highly calibrated gain settings matched to each astronaut. Results from the last two manned Mercury missions produced excellent data.

For the astronaut, biophysical information was largely passive, integrated within the

LEFT A suited Gordon Cooper with the main and left instrument panel segments and a configuration of instrument dials, talkback switches and dials that reveals a layout very typical of aircraft displays of the late 1950s. *(NASA)*

pressure suit itself through appropriate connectors and locator pins described above. The blood pressure measuring system, however, involved an occluded cuff attached to the astronaut's arm, with an external transducer monitoring the differential pressure between the cuff and the cabin pressure. The pulse pressure was a small transducer attached to the astronaut's arm with a pressure source from an oxygen bottle with sufficient quantity for an entire mission. After measuring the blood pressure level the system converted the pressure to a corresponding electrical signal which was then applied to a 2.3kC voltage-controlled oscillator and transmitted by the High and Low telemetry transmitters.

BELOW A view inside spacecraft 18 (Aurora 7) flown as MA-7, the second manned Mercury orbital flight, now preserved at the Museum of Science and Industry, Chicago. *(Via David Baker)*

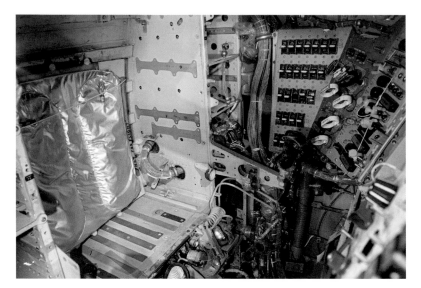

ABOVE The Mercury spacecraft shown here is in the care of the National Air and Space Museum but during 2016 on loan to the National Museum of the US Air Force. It shows the proximity of the left-hand instrument display panel and the fuse panel against the far wall of the spacecraft. *(USAF)*

The blood pressure measurement would be initiated by the astronaut pressing the 'start' switch on the instrument panel, sending a 24V DC pulse for five seconds which caused the system to pressurise the 4.4lb/in^2 (30.3kPa) differential pressure from the oxygen bottle. After pressuring, the system bled off at a linear rate of 0.75lb/in^2 (5.17kPa) in about 22 seconds. The output signal from the pulse sensor was routed through the pressure suit disconnect and over the differential pressure signal. This combined signal was routed through a relay and associated contacts to the two 2.3kC VCOs located in 'D' package and then to the High and Low telemetry transmitters. Sending the signal through the relay was necessary to share the VCOs with the ECG signals.

The first appearance of the signal indicated systolic pressure with a minimum peak amplitude of 150mV while the last occurrence indicated diastolic pressure with a minimum peak of 150mV. The maximum pulse pressure was 1V at peak. On completing the cycle the system remained at rest and when manually initiated by the astronaut the operation continued for one cycle unless manually interrupted with the 'STOP' switch on the main instrument panel.

Connections for the suite of biomedical sensors were originally set on a rectangular board with 16-terminal snaps located to the left of the oxygen inlet port. The relative position of these two devices made it difficult to fasten the connector and repeated use tended to loosen the pins to the point where they were ineffective. A new type of connector was produced by Bendix Aviation Corporation and the location was changed to the outer side of the leg about halfway between the knee and the hip, making it a lot easier to make the connection without damage.

Instruments and indicators

The satellite clock was an electro-mechanical timing device mounted above and to the right of the periscope assembly. It indicated time of day, the elapsed time from lift-off, the time remaining to retrofire and the retrograde time. The time of day was selected by a manually wound spring-driven movement, located in the upper left-hand corner of the clock. Time from launch, time to retrofire and time of firing were indicated on digidial drum counters indicating hours, minutes and seconds for all three functions. Time elements moved in one-step increments and the time to retrofire was supplemented by a telelight situated in the upper right-hand corner of the clock. It would light up five minutes before retrofire and in addition an audible signal could be delivered to the astronaut's helmet ten seconds prior to ignition time.

The clock was automatically started by delivery of a 28V DC signal at lift-off, hence the familiar astronaut call 'And the clock has started', which signalled acceptance that the flight was under way. If this did not occur the astronaut could push a button above but adjacent to the clock to energise it, and the maximum altitude sensor, manually. The time of retrofire was usually calculated before the flight and preset into the clock, but it could be changed manually by the astronaut activating a time-reset handle, or remotely through the command receivers. Two minutes before retrofire the clock would send a signal to the Automatic Stabilization Control System to start the horizon scanners which operated continuously to assure the operation of the rate gyros before the retrograde sequence was initiated. At retrofire a set of contact points within the clock would close, automatically initiating the sequence. Both the time from launch and the retrograde time digidials would be sent to the ground by telemetry.

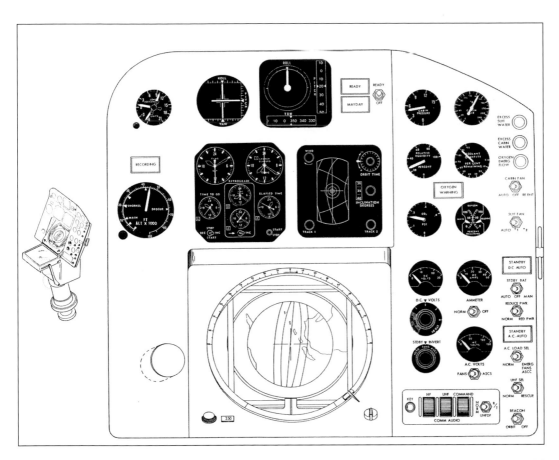

LEFT The main instrument displays changed continuously during the Mercury flight programme, reflecting advancing requirements and specific mission objectives. This early console display does not include the Earth path indicator utilised for orbital capsules. *(McDonnell)*

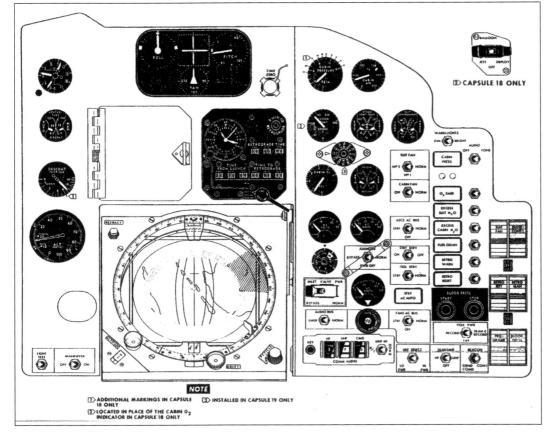

LEFT The main console layout typical of the first three manned flights, with flags noting differences between particular spacecraft. Note that the cover is depicted closed over the Earth path indicator and that spacecraft 19, noted on this diagram, did not fly. *(McDonnell)*

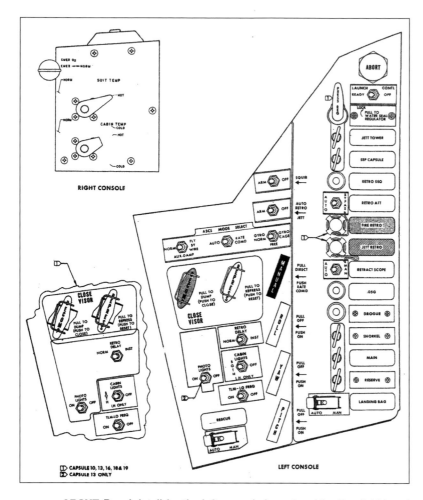

ABOVE Panel detail for the left console layout and for the right hand controls located on the sidewall of the spacecraft to control cabin atmosphere and environment. Flags denote variations between spacecraft.
(McDonnell)

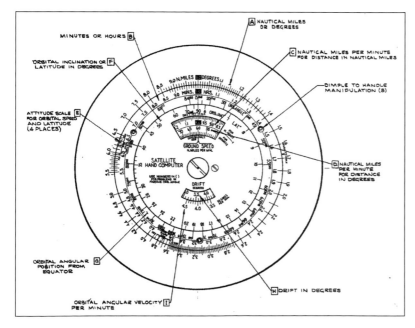

The Earth path indicator consisted of a spherical map that took the form of a globe gimballed and rotating in an appropriate manner to indicate the ground position under the spacecraft. This was a direct transfer from the attitude and position indicator in high-performance aircraft and from the 'eight-ball' fitted in rocket-planes such as the X-15. The near parallel development of the X-15 and the Mercury spacecraft meant that the influence of the NACA/NASA engineers fed back and forth between the two projects, even to the theoretical operation of the control systems.

The Mercury Earth globe was spring-motor powered and was capable of running for 20 hours without rewinding. Approximately 3.85in (9.8cm) in diameter, the globe displayed all continental land-masses, all bodies of water with greater dimensions than 300 miles (483km), the 16 largest rivers of the world and all islands having dimensions of 500 miles (804km). All known island clusters separated from continents by 300 miles and having major axes of less than 500 miles were represented by a 0.20in (0.05cm) diameter dot. The 50 largest cities in the world were represented by a similarly sized dot. Lines of latitude and longitude were scribed on to the globe and numbered in 15° increments.

The controls for the Earth globe were set on the face of the indicator to wind the spring motor and to make adjustments to orbit time, orbit inclination and to slew the globe about the Earth and orbital axes. The splashdown area was displayed as a rectangle and the luminous dot inside the rectangle marked the pre-planned point of impact. The landing area was 3,040 miles (4,891km) ahead of the orbital position above the Earth as indicated by four ring bull's-eyes. This represented the radial distance around the

LEFT Utilised in some respects like a circular slide rule, the hand-held 'computer' enabled the astronaut to take measurements through the periscope and the Earth path indicator and calculate orbital data from distance, altitude and velocity. This opened the door to manual computations for orbital rendezvous developed by astronaut Buzz Aldrin and used to great effect in the Gemini programme. *(McDonnell)*

circumference of the Earth the capsule would travel from retrofire to splashdown. The globe was externally illuminated by the cabin floodlights. Not all missions carried the clock; refer to Chapter 4 for details.

The altimeter reading indicated external pressure above mean sea level and was of the single revolution type, calibrated from 0ft with a calibration marker at 10,000ft (3,048m), the time of parachute deploy, and 20,000ft (6,096m), the height of snorkel deploy. Static pressure was obtained from a centrally located plenum chamber connected to four static ports equally spaced around the small end of the capsule's conical structure.

The longitudinal accelerometer was a completely self-contained unit housed within a hermetically sealed enclosure. It was designed to indicate accelerations in the range of 0g to -9g and 0g to +21g, as measured by accelerators of 32.2ft/sec (9.8m/sec). Attached to the face of the accelerometer were three pointers, one indicating instantaneous acceleration, the remaining two being memory pointers, one of which recorded positive acceleration and the other recording negative accelerations. These memory pointers incorporated a ratchet device that would maintain a deflection until they were reset by means of a knob located on the lower left-hand corner of the accelerometer.

The attitude-rate indicator was a three-axis system located about top-centre of the main instrument panel. It was designed to indicate

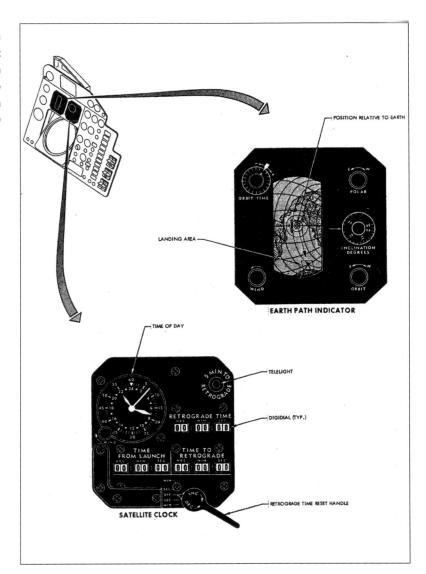

EARTH PATH INDICATOR

SATELLITE CLOCK

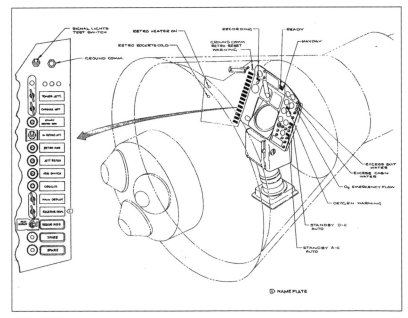

LEFT The Mercury Procedures Trainer in which astronauts rehearsed all the operational phases of a mission and where the concept of the 'simulation supervisor' (known as *simsoup*) arose. A direct lift from the world of aviation, it connected simulated spacecraft controls to an electro-mechanical system controlled by an engineer. Uncompromisingly disposed to throw every possible system failure and anomaly at the occupant to see how he coped with multiple problems, this man was greatly feared for the simulated havoc he could cause! *(NASA)*

BELOW Located to the left of the astronaut's shoulder, the survival kit provided a wide range of items useful for keeping him safe at sea should he splash down far from the recovery forces and have to wait for rescue in a bobbing life raft, itself part of the package. *(McDonnell)*

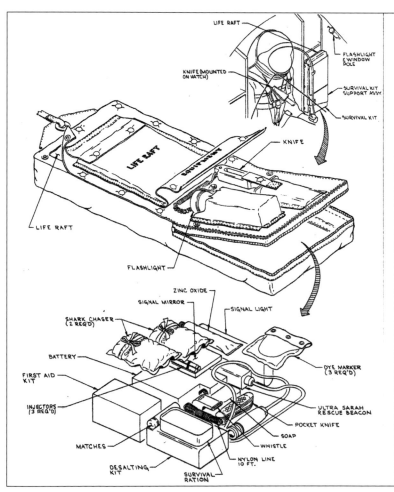

attitude and the attitude change rate and was a composite arrangement of a rate indicator surrounded by a roll attitude indicator, a yaw attitude indicator and a pitch attitude indicator. The rate indicator displayed three pointers, the rate of roll being the pointer that was parallel to the pointer of the roll attitude indicator.

The rates of yaw and pitch pointers were pointing toward the yaw and pitch indicators respectively. Because all components were interchangeable, the replacement of one would not require recalibration of associated components. The indicator was activated by pitch, roll and yaw rate transducers, each one of which comprised a gyroscope, an amplifier and a demodulator. The components functioned together to produce a DC signal proportional to the input rate of change of attitude.

Cameras

A 16mm pulse camera monitoring system was produced by the D.B. Milliken Company of Arcadia, California, and was designed to maintain a complete visual record of astronauts' movements and instrumentation displays. It consisted of an Instrument Observer Camera and a Pilot Observer Camera. The pilot camera was mounted behind the lower left corner of the main instrument panel and was aligned so that it could obtain a view of the astronaut showing upper torso and head. This would be used to map head movements and eyeball focus to determine when he viewed certain instrument displays.

The DBM 7B Instrument Observer Camera had a film capacity of 500ft (152m) and 20,000 frames. It weighed 7.75lb (3.5kg) including film and lens. The DBM 8B Pilot Observer Camera had a capacity of 250ft (76m) and 10,000 frames, weighing 6.75lb (3kg) complete with film and lens. The film selected was a DuPont 16mm double-perforated, high-speed, rapid-reversal type with a Cronar base and had a thickness of 0.0045in (0.0114cm). Each camera had a 10mm f/1.8 Angenieux fixed focus lens with a film-to-lens flange distance modified to allow precise focus at a subject distance of 26in (66cm).

Both cameras were pulse-operated from a programmer with an output of 18–30V and a ten-millisecond input pulse, a cycle being completed in 100 milliseconds. During launch and ascent the film speed was set at three frames per second but during orbital operations the rate was reduced to one frame every ten seconds. The film spools were stacked side-by-side on a coaxial spool shaft with film transport effected by one shuttle claw entering the film simultaneously. Both cameras were provided with an exclusive registration-pin locking device that operated during each exposure to ensure high resolution with elimination of picture-jump or side weave. This locking feature, which Milliken made standard on all its cameras, ensured that each frame was held positively even under conditions of vibration or shock.

The film path was unusual in that the optical axis of the camera was turned 20° to the film spools. As film left the spool it was transported through a 20° twist to the

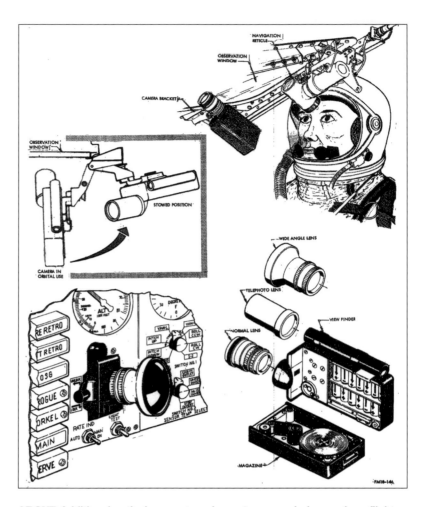

ABOVE Additional optical support equipment was carried on various flights, including a film camera looking directly at the astronaut and another optical device for navigation and visual tracking. *(McDonnell)*

RIGHT The safety of the astronaut was not only the preserve of the spacecraft designer and mission managers. It also extended to the time spent by the astronaut in the launch rocket on the pad. Annotated as Complex 56 on this graphic showing the location of emergency equipment and associated facilities at the Redstone launch site, it was officially the combined launch pads 5 and 6. Work to build pad 5 began in 1954 for routine tests and pad 6 was used for 18 Redstone flights between 20 April 1955 and 27 June 1961. Pad 5 was used for Jupiter-C and Mercury Redstone flights. All operations at the two pads ceased with the launch of MR-4 on 21 July 1961. The complex was handed over to the Air Force Space Museum on 31 January 1964 and is open to the public. *(USAF)*

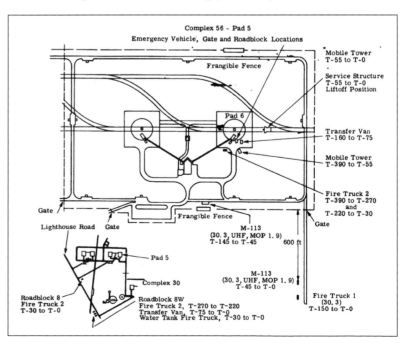

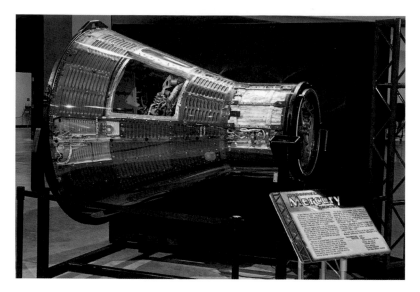

aperture, where it was exposed, then made to climb a spiral loop to the upper level, through a second 20° twist back to normal and finally to the take-up spool. The film was wound so that the emulsion layer was inside on both the feed and take-up spools.

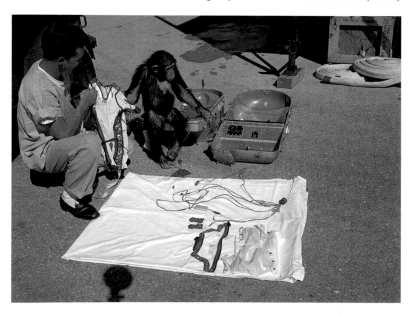

Communications and instrumentation

The instrumentation system provided information on the physical condition of the astronaut and on the condition of the spacecraft and its systems, sent to ground stations through the communications system. The astronaut was provided with a voice communication system active throughout the entire mission, a dual headset and microphone being contained within the helmet and operated through the audio control circuits selected through the voice communications set.

Communications

Voice communication was effected through high-frequency set during launch and on orbit, but after the capsule separated from the launch vehicle adapter communication was only possible through selection by the astronaut. Collins Radio produced the communications equipment for Mercury that was laid out in two functional systems: flight and rescue. Each of these had backup sets and all signals were multiplexed on to a single antenna.

The main HF set was disabled after the antenna fairing was jettisoned after re-entry at which point communications switched to the recovery HF system. Contact via the UHF systems was provided throughout the mission through the Comm UHF equipment and its associated booster amplifier. Transmissions over this set could be made via the UHF switch position on the 'TRANSMIT' selector identical to the main set but without the UHF booster amplifier.

The selected transmitter could be powered up at the operation of a push-to-talk switch or by voice-actuated relay when the 'VOX' switch was in the 'ON' position, at which point the selected transmitter was automatically energised on detection of an audible sound. On splashdown the UHF transmitter would usually be powered-up to provide a direction-finding signal but this automatic feature could be overridden by the astronaut. The command receivers provided an emergency ground-to-capsule voice channel throughout the mission up to splashdown. Power for the communications system was supplied through fuses located in the communications ASCS fuse holders.

Two separate sets of receiver-decoder and auxiliary decoder units were used for reception and decoding of ground command signals, these being used for the purpose of activating various capsule control circuits. Information was picked up throughout the spacecraft in the form of signals from voltage divider circuits. The voltages were modified by coding circuits to enable them to supply suitable inputs to the telemetry transmitters.

Two transmitters were employed for telemetry, each with a power output of 3.3W with their frequencies slightly separated at 228MHz and approximately 260MHz. They were operated continuously from launch until ten minutes prior to splashdown. The power outputs from these transmitters were fed to either the main antenna or the UHF recovery antenna, with power for the system obtained from fuses located in the instrumentation fuse holders.

The beacons in the spacecraft were aids for tracking by ground stations and consisted of C-band and S-band beacons, a UHF recovery beacon powered-up during re-entry, an auxiliary UHF beacon operational at antenna fairing separation and an HF recovery beacon active on splashdown. The beacons provided signals compatible with direction-finding equipment used by the recovery forces. As discussed previously, the UHF voice communication transmitter could be keyed on landing to provide an added signal for direction finding.

The C-band beacon was a transmitter consisting of a receiver and transmitter operating on a frequency of 5,400–5,900MHz, double-pulsed and compatible with FPS-16 radar after ground units had been suitably modified. Input power was from the main pre-splashdown 24V DC bus through the beacon relay controlled by the command receivers, or for continuous operation through the beacon switch. The antenna connection was

through the C-band power divider to the three dedicated and homologated antennas.

The S-band beacon was a transponder operating on a frequency of 2,700–2,900MHz, double-pulsed to reduce the possibility of unauthorised interrogation. This unit was compatible with ground-based Verlort radars and operated at a positive acceptance tolerance of ±0.5 microseconds and a positive tolerance of ±1.8 microseconds. Power circuits, interrogation and replay were the same as for the C-band beacon and antenna connection was through the S-band power divider's three C-band and S-band beacon antennas.

Three C-band and S-band antenna units

BELOW The voice communication system for spacecraft 8 and 9, which flew as MA-3 and MA-5 respectively. Note the instrumentation package. *(McDonnell)*

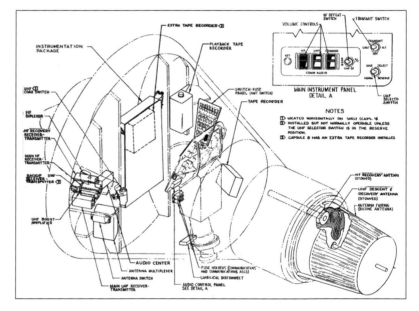

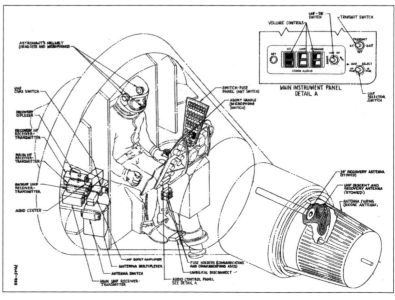

RIGHT For manned flights the configuration shown here was representative for spacecraft 10, 13, 16, 18 and 19 and differed in several ways to the unmanned configurations in the preceding illustration. *(McDonnell)*

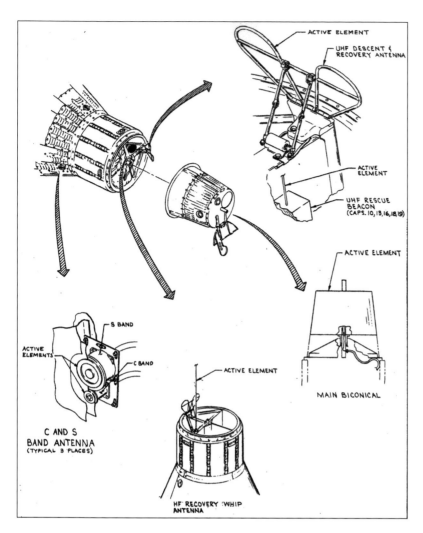

Labels in diagram:
ACTIVE ELEMENT
UHF DESCENT & RECOVERY ANTENNA
ACTIVE ELEMENT
UHF RESCUE BEACON (CAPS. 10,13,16,18,19)
ACTIVE ELEMENT
S BAND
ACTIVE ELEMENTS
C BAND
ACTIVE ELEMENT
MAIN BICONICAL
C AND S BAND ANTENNA (TYPICAL 3 PLACES)
HF RECOVERY WHIP ANTENNA

ABOVE The several antenna requirements were compounded into a single antenna farm housed on top of the recovery compartment and deployed as shown in this diagram.
(McDonnell)

were installed within the capsule structure for the respective beacons with these spaced equally about the circumference of the spacecraft. Each antenna consisted of one helix as a C-band antenna and one as an S-band antenna. Antenna leads were routed through individual power dividers to the three associated helix antennas.

The HF/UHF recovery beacons were combined into a single unit, both energised to provide beacons for recovery equipment. The HF beacon operated on a frequency of 8.364MHz with a tone-modulated output and powered by the 12V standby bus through the impact relay and energised on splashdown. The RF output was fed through the rescue diplexer to the elevated HF recovery antenna. The UHF beacon operated on a frequency of 243MHz with pulse modulation powered by a 6V isolated bus through the antenna fairing separation relay. This circuit was energised during re-entry when the antenna fairing was jettisoned. The RF power

was fed through the antenna multiplexer and the antenna switch to the UHF recovery antenna.

The auxiliary rescue beacon operated at 243MHz with pulse modulation and was also powered by the 6V standby bus through the auxiliary recovery beacon relay which was energised through the antenna fairing separation. The RF power was radiated from the auxiliary rescue beacon antenna and had an output of 91W.

The various types of voice communication, telemetry and beacon receivers and transmitters were served by four antennas. A main UHF antenna was used for the major part of the mission but during re-entry this was jettisoned to allow deployment of the main parachute and replaced by a compact UHF antenna automatically placed in position. On landing the HF recovery antenna was extended. Throughout the mission C-band and S-band antennas were provided for operation of the radar beacons. Antenna switching and multiplexing was carried out by the RF circuitry and power requirements were supplied through a switch-fuse located on the left console panel.

The main biconical HF and UHF antenna was used for prelaunch, launch, orbit and the initial phases of re-entry and was an integral part of the antenna fairing located over the open end of the recovery compartment of the cylindrical afterbody. It served the main HF and UHF voice receiver-transmitters, the command receivers and the telemetry transmitters. The active part of the biconical antenna formed the upper section of the fairing while the lower portion of the fairing and the capsule body formed the ground plane for the antenna.

Employed for the final phase of re-entry, landing and rescue, the compact UHF antenna was located on the open surface of the recovery system's compartment and was folded when the fairing was installed. Sixteen seconds after the fairing was jettisoned the antenna was erected and served the UHF voice receiver-transmitters, the UHF portion of the recovery beacon, the command receivers and the telemetry transmitter. The main UHF voice receiver-transmitter was connected during this period but the radiation from this system was negligible.

The antenna multiplexer enabled simultaneous or individual operation of the radio systems using one antenna. Effectively this was a radio frequency junction box, and final connection to the antenna was through the antenna switch to either the biconical antenna or the UHF recovery antenna. The switch was operated by the antenna fairing separation relay to cause the automatic shift from the main antenna to the UHF recovery antenna at the time of fairing separation. The unit consisted of a number of filters arranged so that all capsule frequencies between 15MHz and 450MHz could be multiplexed on the single feed line, each channel provided with 60db of isolation.

Essentially the capsule was electrically divided in two sections. The antenna fairing structure at the juncture of these sections resembled a discone antenna. This junction was centre-fed by a coaxial cable from the communications sets. At frequencies between 225MHz and 450MHz the antenna fairing acted like a discone antenna. At a frequency of 15MHz it resembled an 'off-centre fed' dipole. Between the upper and lower limits, at 108MHz the unit behaved as a composite dipole-discone antenna. The bicone antenna thus served all frequencies with the exception of C-band and S-band, allowing reception and transmission within the limits of the capsule system.

Contained in a lightweight foam-filled box, the audio centre provided transistorised audio amplifiers, a voice-operated (VOX) relay, an audio filter and tape recorder circuitry and transmitter control circuitry. Two fixed-gain headset amplifiers were used to bring audio signals up to the headset level and feed the headsets separately. Two fixed-gain amplifiers were provided to increase the dynamic microphone output to a level suitable for use with the various transmitters. A low-pass filter with a cut-off at frequencies above 300Hz filtered the audio from the command receivers, enhancing clarity, with outputs from the filter fed to a variable-gain command audio amplifier.

The VOX relay was a transistorised amplifier with separate adjustable threshold level and

RIGHT The onboard tape recorder had a magnesium body for light weight and rigidity and was a crucial element in recording voice traffic on board, as well as the pilot's observations. (NASA)

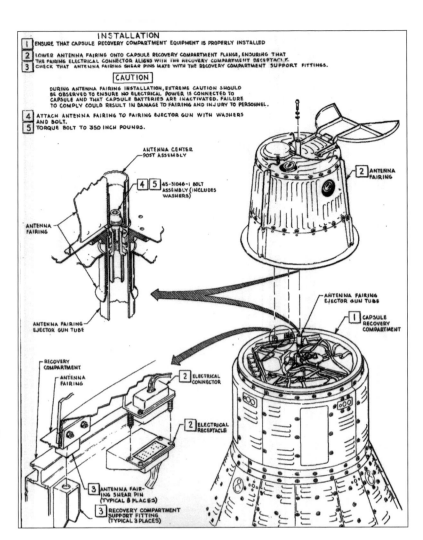

ABOVE An installation diagram from the maintenance manual shows the ejector cartridge used to jettison the fairing on top of the recovery compartment. (McDonnell)

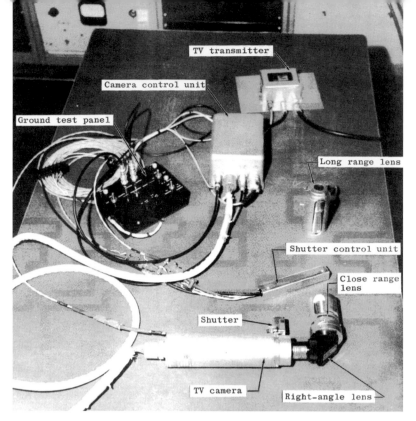

Labels on photo:
- TV transmitter
- Camera control unit
- Ground test panel
- Long range lens
- Shutter control unit
- Close range lens
- Shutter
- TV camera
- Right-angle lens

release time controls operating a relay to provide a grounding circuit for transmitter control. The unit paralleled the external microphone switch, and to operate the relay a three-position VOX switch had to be placed in the 'Trans and Record' position.

A low-power lightweight tape recorder was provided in the spacecraft to provide seven channels of recorded data on 3,600ft (1,109m) of 0.5in (1.3cm) Mylar base tape. The tape transport consisted of a capstan drive supply and take-up reel mechanisms with a DC motor used for power via reduction gearing. A limit switch was provided to interrupt recorder power should the tape break. Record amplifiers were incorporated in the unit for channels 1, 2, 4 and 7, channels 3, 5 and 6 utilising amplifiers incorporated in the commutators located in instrumentation package 'A' in the low-level commutator.

A relay was provided in the audio centre for providing power and signals to the tape recorder. The relay closed a circuit to the tape recorder input so that audio received by the capsule was recorded whenever instrumentation-programmed tape recorder operation was required. Whenever the microphone switch or VOX mode was selected the recorder relay was powered-up. One set of contacts completed the power circuits independent of instrumented programming while a second set routed signals from the microphone amplifiers to the recorder input. The 'Record' position of the VOX switch enabled the astronaut to record his voice without transmitting to the ground.

Instrumentation

The safe operational performance of the Mercury spacecraft and the ability to monitor key functions of its many systems was essential for engineers to understand how the

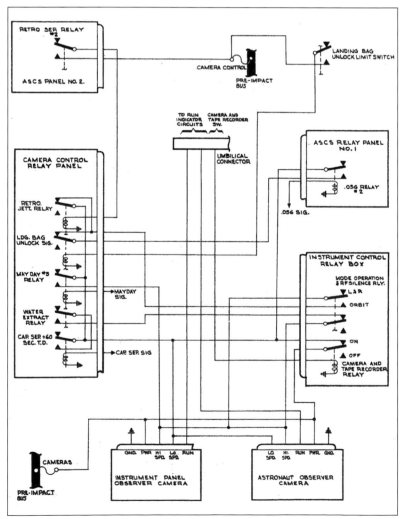

LEFT The circuit diagram for cameras and tape recorder installed in spacecraft 13, 16, 18 and 19. *(McDonnell)*

capsule was performing and why it responded in the way it did to the strenuous environment. The instrumentation section was designed to provide that information and was automatic and semi-automatic in operation from the time power was applied on the launch pad until ten minutes after splashdown. Selective parameters could be controlled or interrogated during flight by either the astronaut or ground command.

Development of the instrumentation system owed much to a combination of flight operations experience with high-speed and high-altitude aircraft carried out by NACA and the Air Force at Edwards Air Force Base, and through experience with instrumenting sounding rockets and probes designed to carry instrumented packages to the fringe of space and transmit data about their performance and the scientific readings they obtained. It was the first integrated effort to advance the state-of-the-art in the collective functions of man and machine and was a monumental contribution to the way all future spacecraft would be instrumented.

The instrumentation system for Mercury comprised various transducers and pickup devices with data coded and applied to the telemetry transmitters and sent to ground stations for immediate analysis and evaluation. In addition the tape recorder was used to capture data for analysis after the flight. The Pilot Observer Camera was also considered part of the instrumentation system so that it could record facial data and register eye alignments. Overall the instrumentation system was planned around three functional groups: monitoring, control, and recording.

Monitoring functions consisted of sampling pressures, temperatures, conditions and the operation of various units and separate functions throughout the capsule. These samples were converted to signals composed of voltages proportional to the conditions in those areas being measured. The proportional voltages were calibrated within common

maximum and minimum ranges to provide zero to full-scale readings. These signals were then channelled into either or both of the two commutators designed for high frequency (HF) and low frequency (LF). These were located in instrumentation package 'A'. Both commutators continuously sampled their respective input channels, combining the signal voltage pulses into a pulse train for each commutator.

Control functions took signals applied to the HF and LF commutators, which were sampled every 0.8 seconds, and placed them in square wave pulses with amplitude between 1V and 3V. These pulses were applied to voltage-controlled oscillators and pulse-duration

BELOW This diagram shows the location of the command receivers for the unmanned missions flown by spacecraft 8 and 9 – MA-3 and MA-5 respectively – which also incorporated an instrumented crewman simulator. *(McDonnell)*

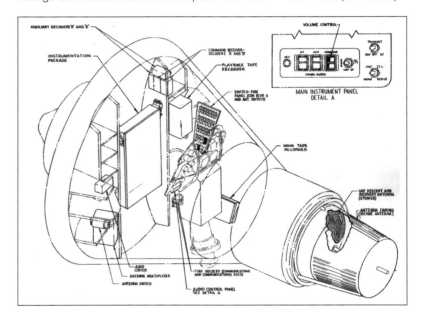

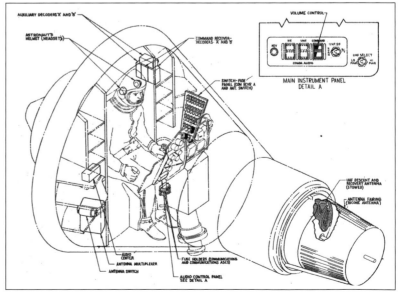

RIGHT As displayed earlier for the antenna and associated control equipment on manned vehicles, spacecraft 10, 13, 16, 18 and 19 had different configurations of equipment for the command receivers. *(McDonnell)*

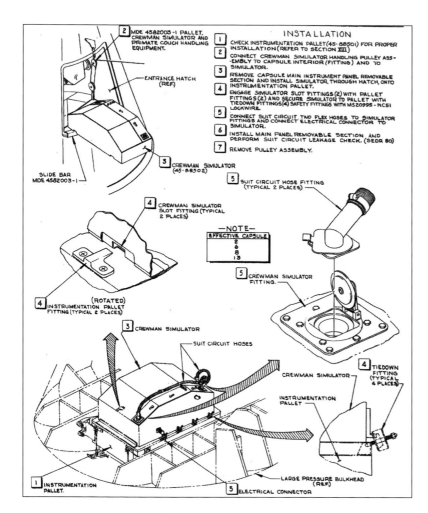

INSTALLATION

2 MDE 4582003-1 PALLET, CREWMAN SIMULATOR AND PRIMATE COUCH HANDLING EQUIPMENT.

1 CHECK INSTRUMENTATION PALLET (45-88501) FOR PROPER INSTALLATION (REFER TO SECTION XIII.)

2 CONNECT CREWMAN SIMULATOR HANDLING PULLEY ASSEMBLY TO CAPSULE INTERIOR (FITTING) AND TO SIMULATOR.

3 REMOVE CAPSULE MAIN INSTRUMENT PANEL REMOVABLE SECTION AND INSTALL SIMULATOR, THROUGH HATCH, ONTO INSTRUMENTATION PALLET.

4 ENGAGE SIMULATOR SLOT FITTINGS (2) WITH PALLET FITTINGS (2) AND SECURE SIMULATOR TO PALLET WITH TIEDOWN FITTINGS (4) SAFETY FITTINGS WITH MS20995 - NCS1 LOCKWIRE.

5 CONNECT SUIT CIRCUIT TWO FLEX HOSES TO SIMULATOR FITTINGS AND CONNECT ELECTRICAL CONNECTOR TO SIMULATOR.

6 INSTALL MAIN PANEL REMOVABLE SECTION AND PERFORM SUIT CIRCUIT LEAKAGE CHECK. (SEDR 80)

7 REMOVE PULLEY ASSEMBLY.

ENTRANCE HATCH (REF.)

SLIDE BAR MDE 4582003-1

3 CREWMAN SIMULATOR (45-88502)

4 CREWMAN SIMULATOR SLOT FITTING (TYPICAL 2 PLACES)

5 SUIT CIRCUIT HOSE FITTING (TYPICAL 2 PLACES)

—NOTE—
EFFECTIVE CAPSULE
2
6
8
13

5 CREWMAN SIMULATOR FITTING.

4 (ROTATED)
4 INSTRUMENTATION PALLET FITTING (TYPICAL 2 PLACES)

3 CREWMAN SIMULATOR

SUIT CIRCUIT HOSES

CREWMAN SIMULATOR

INSTRUMENTATION PALLET

4 TIEDOWN FITTING (TYPICAL 4 PLACES)

1 INSTRUMENTATION PALLET.

LARGE PRESSURE BULKHEAD (REF.)

5 ELECTRICAL CONNECTOR

LEFT The crewman simulator was specifically configured to plug into systems appropriate for doubling up in place of a human occupant, providing a wide range of interactions with spacecraft systems. *(McDonnell)*

modulation converters. The frequencies of the oscillators varied between 10.5KHz, ±6.3/4%, by the commutated pulse amplitude signals. The frequency-modulated outputs of the 10.5KHz voltage-controlled oscillators were applied to two mixers. The commutator pulse amplitude modulation signals were also applied to pulse duration modulation converters, which re-formed the wave-shapes to obtain pulse duration wave trains. These trains were then applied to the tape recorder.

Amplified aeromedical signals were coupled in pairs of 1.3KHz, 1.7KHz and 2.3KHz voltage-controlled oscillators. A zero to full-scale signal caused a deviation in centre frequency of ±6.3/4%. The output of the oscillators was applied to the mixers where the commutated outputs and the aeromedical signals were combined. Mixer 'A' accepted a signal from the compensating oscillator which served as a reference during data evaluation to indicate

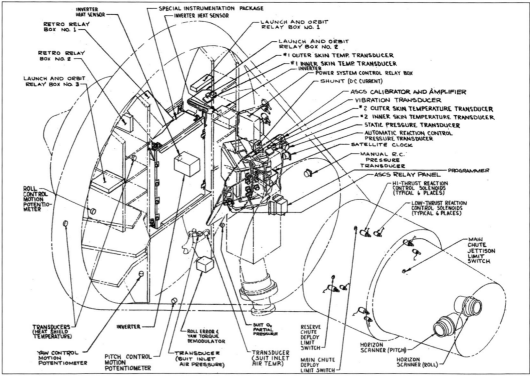

RIGHT The instrumentation suite for spacecraft 8 contained a very broad and diverse range of transducers, relays and sensing and monitoring switches for the unmanned flight of spacecraft 8 on the MA-3 mission. *(McDonnell)*

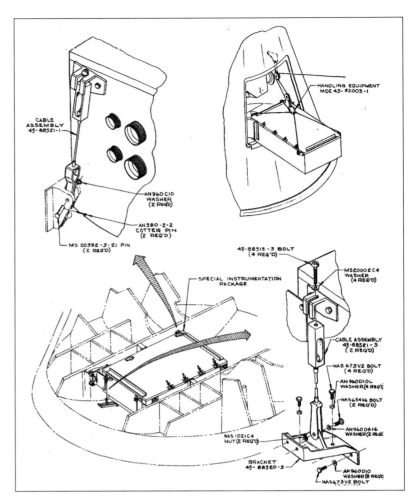

fluctuations in tape speed. The composite signal from each mixer was applied to the tape recorder, the ground-test umbilical and a telemetry transmitter.

Recording functions were related to the tape recorder and the Pilot Observer Camera. The tape recorder provided a full recording time of 6.4 hours but this was extended for MA-9 (see 'Flight Results', starting on page 163).

Before flight, ground-testing and control of the instrumentation package was provided through the umbilical receptacles where non-radiating checks could be performed through the direct telemetry link. The instrumentation system controlled power to its own and other systems through mode relays and a programmer. The water extractor in the environmental control system was also programmed at regular intervals during a typical mission. The system also programmed

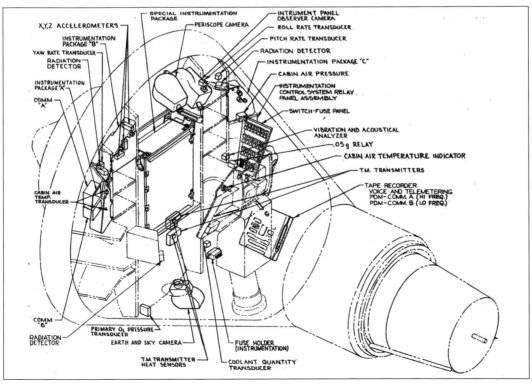

RIGHT In comparison to the unmanned flight of spacecraft 8, the objective for spacecraft 9 was to fly chimpanzee Enos on MA-5. Here the capsule containing the chimpanzee had its own dedicated instrumentation pallet. *(McDonnell)*

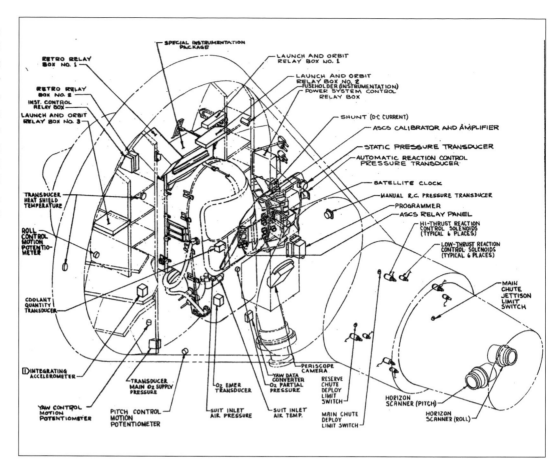

RIGHT As with the unmanned flight of spacecraft 8, spacecraft 9 also had dedicated camera and recording equipment to obtain information relevant to this simulation of a human flight. *(McDonnell)*

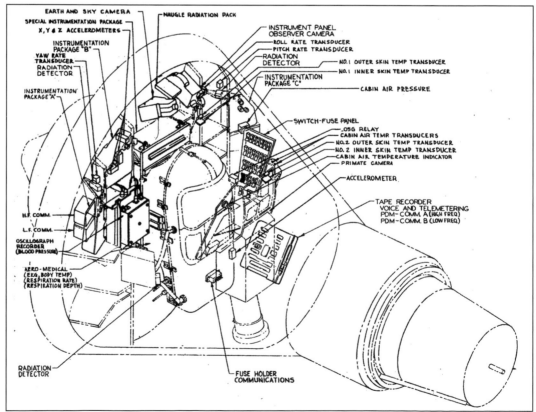

operation of the Pilot Observer Camera at both high (6fps) and low (5fps) frame rates according to the acuteness of the requirement.

Calibration voltages for maximum readings were supplied periodically to the monitoring instrumentation circuits and this was also done before launch by ground command and at intervals during orbital operations. Potentiometer-type transducers were connected across the instrumentation with 3V DC power, and control-stick potentiometers were provided on the basis of one per axis. The satellite clock, manual and automatic helium supply pressures, attitude indicators for the ASCS, main and reserve oxygen pressure sensors, static pressure, suit pressure and coolant quantity pressure readings were also read by potentiometers.

A special instrumentation pallet was developed for the primate launches flown as MR-2 and MA-5 with aeromedical instrumentation the same as that for flights with humans except for different calibrations for ECG and respiration values. Only three ECG transducers were attached to the primate. The primate's couch was mounted to a special pallet and instrumented its reactions and responses. The primate programmer consisted of an electronic panel assembly and a liquid feeder. The specifics of the apparatus for the two primate flights were different for the ballistic and orbital missions and were the responsibility of McDonnell and the US Air Force 6571st Aeromedical Research Laboratory.

The ballistic MR-2 flight was flown by a male chimpanzee named Ham, weighing 37lb (17kg), who had received 219 hours' familiarity over a 15-month period. The performance test panel he used contained three lights and two levers, each 1in (2.54cm) in diameter and protruding 2.25in (5.7cm). To prevent launch accelerations from activating it, a force of 2lb (0.9kg) was required to move the lever. Both suborbital and orbital configurations followed the same protocol in that the tests were made to ascertain whether the restrained animal could respond normally during weightlessness to a series of problems graded in difficulty and in motivation.

The orbital flight of the 42lb (19kg) male chimpanzee Enos on the MA-5 flight involved a test panel with three levers, the inline digital displays being spaced 4in (10cm) above the bottom of the panel. The centre display was mounted in the midpoint of this line and the two additional displays were on centrelines 3.8in (10cm) to the left and right of this point. As with the MR-2 panel, a 1in (2.54cm) diameter lever was centred and mounted below each display, each protruding 2.25in (5.7cm) beyond the face of the panel, and required a 2lb (0.9kg) force for application.

A pellet feeder was also incorporated, consisting of a tube which served 106 specially designed food pellets weighing 0.5gm each. A spring maintained pressure on the single column of pellets in a magazine, or tube, dispensing one at a time, delivered by a solenoid that pushed the pellets from the tube into two small plastic fingers located on the face of the panel below and to the right of lever 3. A lip-lever drinking tube was mounted in the couch to the right of the subject's head, and the illumination of a small green lamp mounted either above or below this tube served to cue the chimpanzee that water was available and he could bite the lever.

The method of programming the tasks, the procedure for delivering a mild electric shock and the shock level were the same for both ballistic and orbital configurations. In one, data was routed through four points on a multiplex

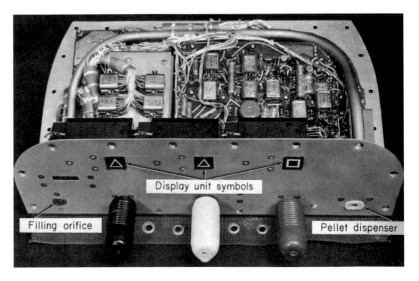

BELOW The instrumentation pallet for Enos contained a special set of tasks that required levers to be pulled for a reward, this being an integral part of the instrumentation delivered to the ground. *(NASA)*

RIGHT The pallet for Enos carried wire carry-through ducts to the spacecraft's instrumentation suite for sending signals to Earth via the telemetry links. (NASA)

FAR RIGHT Part of the bioinstrumentation harness for chimpanzee Ham, who would fly the ballistic MR-2 mission. (NASA)

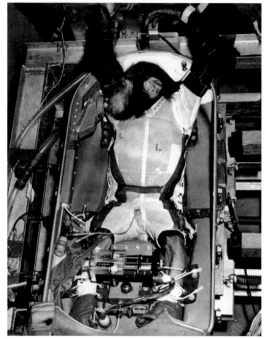

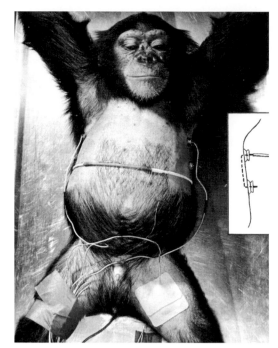

BELOW The flight couch for Ham, with associated garments, biosensors and personalised medical suit. (NASA)

unit and the commutated data was transmitted to the ground. The data were also retained on the tape recorder (see pages 155–9). In the second the performance was coded by voltage-controlled oscillators and continuously recorded on two sub-channels of the onboard recorder tape. A single digital counter was mounted behind the performance panel and served as a final method for collecting the performance data.

The ballistic programme posed two shock-avoidance tests where the chimpanzee had to depress levers in response to coloured light stimuli within a certain time period to avoid a mild shock. The orbital unit had three additional programmes during various phases of which the primate obtained drink and banana-flavoured pellet rewards or shock punishment for error or failure to respond. The stimuli for the particular programme consisted of coloured lights and symbols in display units.

The instrument panel contained three display units. The left display had one consisting of six symbols and a cyan blue disc display. The centre display had six symbols, a yellow disc and a white disc display. The right-hand panel had six symbols, a red disc and a green display. Three switch levers were provided for the primate to operate, each switch closure being recorded on a four-digit counter mounted near the left switch lever. The pellet dispenser would dispense food on command of a 28V pulse. The liquid dispenser would also deliver a drink of liquid when actuated by a 28V pulse and triggered by the animal's lips. Power for the couch was supplied from a dedicated instrumentation fuse block through the special instrumentation relay.

The animal's responses during various programmed sequences were recorded and telemetered to the ground. Psychomotor tests included a series of evaluations including 15 minutes of regular and classified shock avoidance, two minutes of timed-out activities,

Restraint straps

Bioinstrumentation cable
ECG, respiration, body
temperature, and shock
electrodes

Suit

Bib

Waterproof
pants

five minutes of ration reward, 18 combinations of symbols in five minutes and 18 combinations of '1' and 'X' symbols. If the last set of 36 combinations were completed before the elapsed five minutes was over the entire sequence was reset and started over. The test sequences were a complex interactive set of actions, prompts and responses in an increasingly complex set of activities.

Flight results

The instrumentation system monitored more than 100 parameters during a typical mission and provided vital data that validated the Mercury programme requirement for a comprehensive evaluation of the environment and the performance of systems and subsystems that, in their design, had been only a best-fit from an uncertain database into an unexplored environment. The basic purpose in sending astronauts into space during the Mercury programme was to characterise the weightless environment of space and to judge how well man and machine could perform.

Continuous analysis and re-evaluation of what systems were essential and which could be discarded, to save weight or to reduce complexity, was maintained throughout the flight programme. The decision to eliminate two redundant telemetry transmitters on the last manned MA-9 mission was a direct result of continuous evaluation on a learning curve that grew quickly over the 25 months of manned operations.

Two classic examples suffice to demonstrate how valuable the instrumentation system was in providing legacy examples of what was right and what was wrong with the instrumentation system. The first three orbital missions provided adequate baseline data but there was a requirement for more data storage capacity, and that was sought without incurring additional weight; it became essential to record much more data on the 22-orbit flight of MA-9, which would accumulate on that one flight much more than the cumulative total of all five preceding manned space flights. To do that the tape speed was reduced from 1.875in/sec (4.76cm/sec) to 0.937in/sec (2.38cm/sec) and the

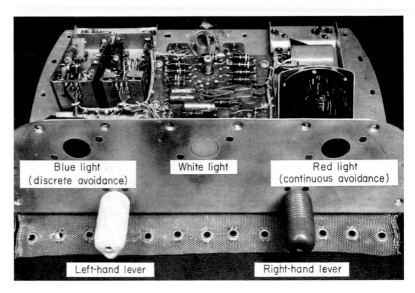

ABOVE Due to its much shorter duration, the activation pallet for experimental tasks – pre-programmed to evaluate Ham's performance in space on the MR-2 flight – was a simplified version of that used for Enos on the Atlas flight. *(NASA)*

periods during which data was accumulated were reprogrammed.

The other example demonstrated how the need for data could work against the reliability of a mission. When John Glenn returned to Earth at the end of his three-orbit flight a telemetry signal indicated that the heat shield had unlatched from the forebody of his spacecraft. Ground controllers believed this to be an incorrect indication and requested that he leave the retro-package on during re-entry so that the straps would hold the shield secure against its mounting until the atmosphere pressed against the capsule and held it tight by pressure alone and the retro-package burned away. It was believed by flight controllers that the telemetry signal was incorrect and that proved to be the case.

Subsequent detailed analysis of the spacecraft and the various switches revealed that a bent and loose shaft on the limit switch had registered an erroneous signal and that movement of the sensor would send such a signal without displacing the shaft. Various procedures were examined in strenuous efforts to track down why this quality control issue had occurred and changes were made to the installation procedures.

In other applications of instrumentation data,

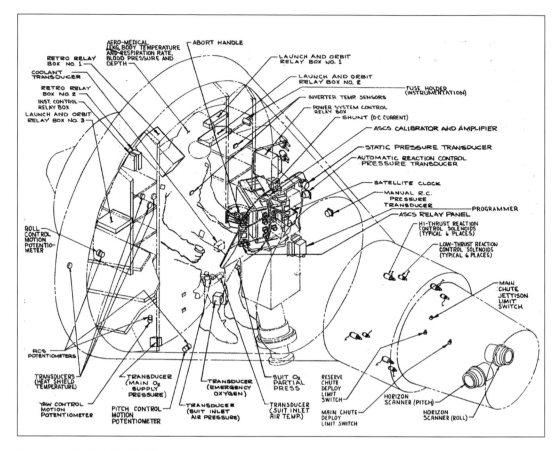

RIGHT Laid out on this schematic, the instrumentation configuration for manned missions was precisely aligned with the mission objectives on the MA-7 and MA-8 flights. *(McDonnell)*

Labels (clockwise): AERO-MEDICAL EKG, BODY TEMPERATURE AND RESPIRATION RATE, BLOOD PRESSURE AND DEPTH; ABORT HANDLE; RETRO RELAY BOX NO. 1; COOLANT TRANSDUCER; RETRO RELAY BOX NO. 2; INST. CONTROL RELAY BOX; LAUNCH AND ORBIT RELAY BOX NO. 3; LAUNCH AND ORBIT RELAY BOX NO. 1; LAUNCH AND ORBIT RELAY BOX NO. 2; FUSE HOLDER (INSTRUMENTATION); INVERTER TEMP. SENSORS; POWER SYSTEM CONTROL RELAY BOX; SHUNT (D-C CURRENT); ASCS CALIBRATOR AND AMPLIFIER; STATIC PRESSURE TRANSDUCER; AUTOMATIC REACTION CONTROL PRESSURE TRANSDUCER; SATELLITE CLOCK; MANUAL R.C. PRESSURE TRANSDUCER; PROGRAMMER; ASCS RELAY PANEL; HI-THRUST REACTION CONTROL SOLENOIDS (TYPICAL 6 PLACES); LOW-THRUST REACTION CONTROL SOLENOIDS (TYPICAL 6 PLACES); MAIN CHUTE JETTISON LIMIT SWITCH; HORIZON SCANNER (PITCH); HORIZON SCANNER (ROLL); RESERVE CHUTE DEPLOY LIMIT SWITCH; MAIN CHUTE DEPLOY LIMIT SWITCH; TRANSDUCER (SUIT INLET AIR TEMP.); SUIT O₂ PARTIAL PRESS; TRANSDUCER (EMERGENCY OXYGEN); TRANSDUCER (SUIT INLET AIR PRESSURE); PITCH CONTROL MOTION POTENTIOMETER; TRANSDUCER (MAIN O₂ SUPPLY PRESSURE); YAW CONTROL MOTION POTENTIOMETER; TRANSDUCERS (HEAT SHIELD TEMPERATURE); RCS POTENTIOMETERS; ROLL CONTROL MOTION POTENTIOMETER

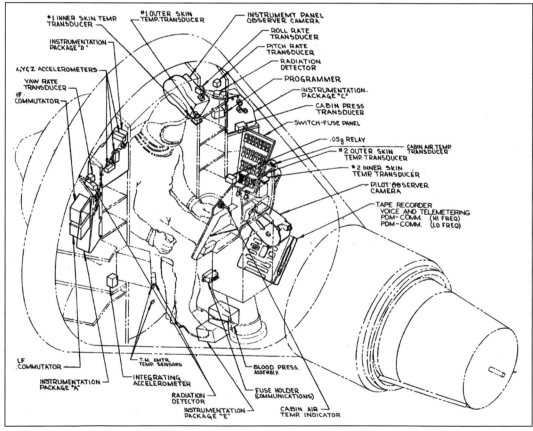

RIGHT Connections to instrument package 'A' are displayed on this layout map of sensors, including biomedical data obtained from the pilots on MA-7 and MA-8. *(McDonnell)*

Labels: #1 INNER SKIN TEMP. TRANSDUCER; #1 OUTER SKIN TEMP. TRANSDUCER; INSTRUMENT PANEL OBSERVER CAMERA; ROLL RATE TRANSDUCER; PITCH RATE TRANSDUCER; RADIATION DETECTOR; PROGRAMMER; INSTRUMENTATION PACKAGE "C"; CABIN PRESS TRANSDUCER; SWITCH-FUSE PANEL; .05g RELAY; CABIN AIR TEMP TRANSDUCER; #2 OUTER SKIN TEMP TRANSDUCER; #2 INNER SKIN TEMP TRANSDUCER; PILOT OBSERVER CAMERA; TAPE RECORDER VOICE AND TELEMETERING PDM-COMM. (HI FREQ) PDM-COMM. (LO FREQ); BLOOD PRESS. ASSEMBLY; FUSE HOLDER (COMMUNICATIONS); CABIN AIR TEMP. INDICATOR; INSTRUMENTATION PACKAGE "E"; RADIATION DETECTOR; INTEGRATING ACCELEROMETER; T.M. XMTR. TEMP. SENSORS; INSTRUMENTATION PACKAGE "A"; LF COMMUTATOR; IF COMMUTATOR; YAW RATE TRANSDUCER; X,Y&Z ACCELEROMETERS; INSTRUMENTATION PACKAGE "D"

RIGHT Engineers at NASA's Lewis Flight Research Center evaluate the servo instrumentation for the Mercury spacecraft. *(NASA)*

in-flight monitoring of systems performance allowed some changes in the equipment carried on successive missions. One of the two command decoders and the HF recovery transceiver were removed for the MA-8 and MA-9 flights while one of the UHF telemetry transmitters was also deleted from the instrumentation system on MA-9. Some data, however, led to changes as a result of insufficient data prior to operational experience. On the first five manned missions the HF recovery beacon proved virtually ineffective and only by analysing the way the whip antenna deployed could changes be made in time for the MA-9 flight.

It appeared that pyrotechnically deploying the antenna at splashdown caused it to come out while covered with water, and the combination of the electrically conductive products of combustion from the explosive charge plus water contamination caused the problem. For MA-9 the explosive cartridge was replaced with a nitrogen pressurisation system that was nonconductive.

Several changes and modifications were made to the attitude control systems where on MA-4 an open circuit in the pitch-rate gyro input to the amplifier-calibrator caused spacecraft attitude to be in error at retrofire, resulting in it landing 86 miles (138km) off target. There were few malfunctions to the complex ASCS or the RSCS systems as measured by the instrumentation on board although a short circuit in two of the power plugs to the autopilot on MA-9 caused loss of automatic responses on the last few orbits.

CENTRE The telemetry suite for spacecraft 8, flown unmanned and carrying a dedicated instrumentation package. *(McDonnell)*

RIGHT The equivalent telemetry system installed in the manned orbital spacecraft, in this illustration specifically for spacecraft 10, 13, 16, 18 and 19. *(McDonnell)*

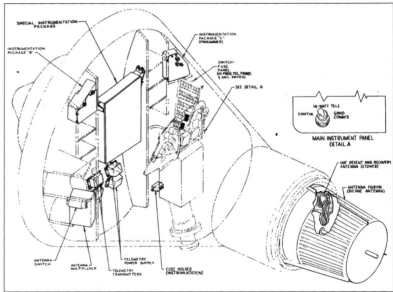

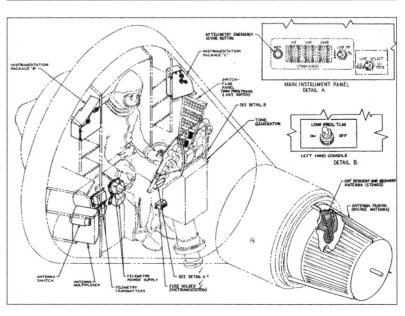

THE MERCURY SPACECRAFT (1959–63)

Chapter Four

Performance analysis (1959–63)

NASA supported 25 flight events involving several different launch vehicles and a variety of spacecraft from boilerplates to qualification capsules and six manned space missions. They were the first of their kind, without parallel and devoid of the benefit of precedence from which to learn. True pioneers.

In this chapter the technical changes to each spacecraft are explored and essential flight performance assessed. It does not include detailed accounts of the missions themselves as this book is purely an engineering evaluation and assessment.

OPPOSITE John Glenn manoeuvres himself into spacecraft 13 (*Friendship 7*) for the first manned US orbital flight on 20 February 1962. *(NASA)*

ABOVE Big Joe, an Atlas rocket carrying a boilerplate Mercury spacecraft with an escape system, was launched on 9 September 1959. *(NASA)*

BELOW Military personnel prepare a monkey for flight on a Little Joe rocket to test the launch escape system. *(USAF)*

Flight events

Little Joe-1/boilerplate
Launched: 21 August 1959
Flight duration: 0hr 0min 20sec

Planned as a test of the launch escape system at altitude, it failed when activation of the LES occurred prematurely after a ground potential was supplied by a battery being placed on charge, causing a transient current to trigger the LES 35 minutes before the planned lift-off time. Leaving the LJ-1 launcher on its pad at Wallops Island, Virginia, the capsule shot to a height of 2,100ft (640m), separated the tower, deployed the drogue but had insufficient electrical power to deploy the main parachute. The boilerplate was destroyed on impact.

Big Joe/boilerplate
Launched: 9 September 1959
Flight duration: 0hr 13min 00sec

Test of the ablation shield on Atlas 10-D to a height of 95 miles (153km), a range of 1,496 miles (2,407km), a speed of 14,487mph (23,905kph) and a maximum aerodynamic pressure (max-q) of 675lb/in² (4,654kPa). No launch escape system. Trajectory compromised when the booster engines failed to separate, putting the spacecraft on the water 500 miles (804km) short. Boilerplate was retrieved seven hours later.

Little Joe-6/boilerplate (unmanned)
Launched: 4 October 1959
Flight duration: 0hr 5min 10sec

Utilising the same rocket as Little Joe-1, the boilerplate was fired with an inert launch escape tower to a height of 37 miles (59.5km), a range of 79 miles (127km) and a speed of 3,075mph (4,948kph) to test the dynamic environment. Max-q of 3,400lb/in² (23,443kPa) was the highest experienced on any Mercury flight. The Little Joe was intentionally destroyed to prove the integrity of the destruct system.

Little Joe-1A/boilerplate (unmanned)
Launched: 4 November 1959
Flight duration: 0hr 8min 11sec

Max-Q and launch escape tower test achieved an altitude of 9 miles (14.5km), a range of 11 miles

(17.7km) and a speed of 2,022mph (3,253kph). Max q of 168lb/in² (1,158kPa) and a g load of 16.9, the highest recorded on any Mercury flight. Slow ignition of the escape rocket did not conform to the trajectory required for the test objectives but the spacecraft was recovered safely and used again on the MR-BD flight of 24 March 1961.

Little Joe-2/boilerplate (unmanned)
Launched: 4 December 1959
Flight duration: 0hr 11min 06sec
High-altitude abort test with Rhesus monkey 'Sam' initiated at a height of almost 19 miles (30.6km), maximum height achieved 53 miles (85km), range of 194 miles (312km), speed of 4,466mph (7,186kph), max-q of 2,150lb/in² (14,824kPa), second highest on any Mercury flight. Monkey safely recovered.

Little Joe-1B/boilerplate (unmanned)
Launched: 21 January 1960
Flight duration: 0hr 8min 35sec
Max-Q abort test with monkey 'Miss Sam' to replace failed flights LJ-1 and LJ-1A, achieved a height of 9 miles (14.5km), a range of 12 miles (19km), a speed of 2,022mph (3,253kph) and a max-q of 1,070lb/in² (7,377kPa). Monkey recovered safely.

Beach Abort/spacecraft 1 (unmanned)
Launched: 9 May 1960
Flight duration: 0hr 1min 16sec
First major qualification flight with a production capsule launched off a fixed pad from Wallops Island by the launch escape tower on a simulated pad abort. Maximum height of 2,600ft (792m), range of 1 mile (1.6km), maximum speed of 976mph (1,570kph). Spacecraft 1 was outfitted with all major systems and qualified for pad-abort operations in the event of a malfunction to a planned suborbital or orbital launch.

LEFT Two Little Joe flights in December 1959 and January 1960 were launched to test the escape system and its effect on monkeys. The first launch (LJ-2) involved a recovery by the US Navy 194 miles offshore. *(USN)*

RIGHT Engineers prepare spacecraft 1 for the one and only beach abort test to evaluate the performance of the escape system from ground level, simulating a capsule fired off the top of a rocket still on the launch pad. *(NASA)*

RIGHT On 9 May 1960 the beach abort test fired spacecraft 1, successfully demonstrating recovery of the boilerplate little more than one minute later. The jettison rocket is fired in images 4 and 5 and the separated tower is falling free in images 6, 7 and 8.
(NASA)

Mercury Atlas-1/spacecraft 4 (unmanned)

Launched: 29 July 1960
Flight duration: 0hr 3min 18sec

No escape system, test of structural integrity of production spacecraft during re-entry from altitude but the mission terminated at 60 seconds when the adapter interface failed, buckling the top of the Atlas 50-D. Maximum height reached was 8.1 miles (13km), range 6 miles (9.6km), speed 1,701mph (2,737kph). Modifications were made to the adapter and the top end of the Atlas launch vehicle.

Little Joe-5/spacecraft 3 (unmanned)

Launched: 8 November 1960
Flight duration: 0hr 2min 22sec

Complete production spacecraft without landing bag, partially operational ASCS, partial communications system, in an abort test but the escape tower fired early, the spacecraft failed to separate and all hardware destroyed on impact with the sea. Maximum altitude of 10.1 miles (16.25km), range at impact of 14 miles (22.5km), speed 1,785mph (2,550kph). Modifications were made to subsequent spacecraft to prevent premature firing of escape motor.

Mercury Redstone-1/spacecraft 2 (unmanned)

Launched: 21 November 1960
Flight duration: 0hr 0min 02sec

Attempted qualification flight for Mercury Redstone launcher (MR-1) on a suborbital flight and test for flight-rated spacecraft, failed when rocket motor shut down after lifting off 3.8in (9.6cm) and settling back down on its launch

RIGHT Emerging from a crate stamped 'FRAGILE – HANDLE WITH CARE', the flight couch for chimpanzee Ham is given a checkout by the space-age ape before his flight on MR-2.
(NASA)

pedestal. A sneak circuit through the control plug and ground network gave a premature booster cut-off signal. The refurbished spacecraft would fly as MR-1A a month later.

Mercury Redstone-1A/ spacecraft 2A (unmanned)
Launched: 19 December 1960
Flight duration: 0hr 15min 45sec
Repeat of the MR-1 flight a month earlier with all flight objectives met using MR-3 launch vehicle, despite a higher velocity than planned at cut-off, achieving 4,909mph (7,898kph), an altitude of 130.7 miles (210km), a range of 235 miles (378km) and a higher than planned g-load of 12.4. Mission objectives were met and spacecraft operated satisfactorily.

Mercury Redstone-2/spacecraft 5 (unmanned)
Launched: 31 January 1961
Flight duration: 0hr 16min 39sec
Carried chimpanzee 'Ham' on launch vehicle MR-2 on a ballistic suborbital qualification shot to demonstrate the Mercury Redstone was viable for human flight on a similar trajectory. An early cut-off due to premature fuel depletion caused the escape tower to fire, taking the spacecraft and its occupant to an abort, an altitude of 157 miles (252km), a speed of 5,857mph (9,424kph) and a range of 422 miles (679km), about 126 miles

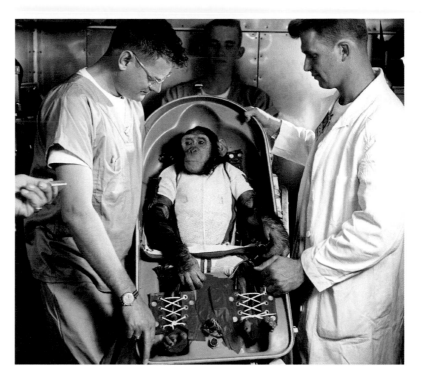

ABOVE The two chimpanzees used by NASA in Mercury tests were well trained, had no physical adversity or trauma and were relaxed and eager to learn how to operate controls for rewards. Here Ham checks out his flight canister. *(NASA)*

BELOW LEFT Ham relaxes after his historic fight to a higher altitude than expected. *(USN)*

BELOW Ham sits it out on an Air Force truck, used to delays, postponements and countdown holds just like human astronauts! *(NASA)*

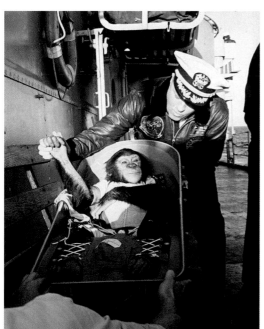

(202km) further downrange than planned, causing a 2hr 56min delay in recovering the capsule and its occupant. Several modifications were made to the spacecraft as a result of this flight, including an additional fibreglass bulkhead between the large pressure bulkhead and the heat shield, a modified heat shield retention system and improvements to the inflow valve.

Mercury Atlas-2/spacecraft 6 (unmanned)
Launched: 21 February 1961
Flight duration: 0hr 17min 56sec

Atlas 67-D launched on a repeat of MA-1 test objectives with several structural modifications made to the adapter, the top of the Atlas and an 8in (20cm) wide stainless steel band around the upper end of the liquid oxygen tank. The primary objective was to expose the capsule to a severe re-entry trajectory and evaluate loads and temperatures imposed. Maximum altitude reached was 114 miles (183km), range 1,432 miles (2,301km), speed 13,227mph (21,282kph) with a max-q of 991lb/in² (6,833kPa) and a g-load of 15.9. All test objectives were met.

Little Joe-5A/spacecraft 14 (unmanned)
Launched: 18 March 1961
Flight duration: 0hr 23min 48sec

Added to the flight-test schedule after the failure of LJ-5, the rocket carried a spacecraft equipped only with systems relevant to this flight. Again, early ignition of the escape rocket occurred but this time a ground command resulted in separation of the capsule from Little Joe. Due to excessive dynamics loads both parachutes deployed simultaneously but a safe landing resulted from both filling with air. Changes to Little Joe were made to reduce air-loads in the area by an improved fairing on the booster rocket. Maximum height reached was 7.7 miles (12.4km), range 18 miles (29km) and speed 1,783mph (2,869kph) for a max-q of 1,580lb/in² (10,894kPa) and 8g.

Mercury Redstone-booster development/boilerplate (unmanned)
Launched: 24 March 1961
Flight duration: 0hr 8min 23sec

Carried the boilerplate previously mated to

the Little Joe-1A rocket in a test of a series of modifications to the Redstone rocket that exhibited technical problems on MR-1A and MR-2 flights. Maximum height attained was 113.5 miles (182.6km), range 307 miles (494km) and speed 5,123mph (8,243kph) for a max-q of 580lb/in^2 (3,999kPa).

Mercury Atlas-3/spacecraft 8 (unmanned)

Launched: 25 March 1961
Flight duration: 0hr 7min 19sec

Atlas 100-D was assigned to carry the spacecraft into orbit for one full revolution of the Earth as a qualification flight of launcher and capsule but the flight was aborted at 40 seconds when it failed to roll and pitch to the appropriate azimuth due to an electrical malfunction in the Atlas autopilot. Maximum altitude was 4.5 miles (7.2km) and speed 1,177mph (1,894kph) but range zero due to vertical climb. Exposed the capsule to max-q of 880lb/in^2 (6,067kPa) and 11g, from which it was successfully recovered after an abort off the malfunctioning Atlas.

Little Joe-5B/spacecraft 14A (unmanned)

Launched: 28 April 1961
Flight duration: 0hr 5min 25sec

Carried the refurbished spacecraft from LJ-5A equipped only with mission-specific systems, no landing bag and several other subsystems. Limit switch modifications were fitted to the spacecraft-adapter clamp ring that would prevent the escape system firing until the band separation bolts had fired. Delayed ignition on two Little Joe rockets significantly changed the trajectory, imposing a max-q of 1,920lb/in^2 (13,238kPa), instead of a planned 900lb/in^2 (6,205kPa), and 10g. Altitude reached was 2.8 miles (4.5km), speed 1,780mph (2,864km) and range 9 miles (14.5km).

Mercury Redstone-3/spacecraft 7 *Freedom 7* (Alan Shepard)

Launched: 5 May 1961
Flight duration: 0hr 15min 28sec

The first manned flight of a Mercury spacecraft on a suborbital flight by MR-7 launch vehicle reaching an altitude of 116.5 miles (187km),

LEFT Alan B. Shepard launches in spacecraft 7 (*Freedom 7*) on a ballistic shot (MR-3) on 5 May 1961, America's first manned space flight. *(NASA)*

a range of 303 miles (487km) and a speed of 5,134mph (8,260kph), a max-q of 580lb/in^2 (3,999kPa) and maximum acceleration of 11g, enduring 5min 16sec of weightlessness. Shepard experimented with the attitude thrusters in manual control on an axis-by-axis basis and became overcrowded with tasks. Successful flight, no significant anomalies, pilot electing to emerge from the side hatch, retrieval by the carrier USS *Lake Champlain*. This flight had considerable significance. Watched by President John F. Kennedy from the White

BELOW Shepard is recovered by the USS *Lake Champlain*, the pilot and the capsule winched aboard separately by helicopter. *(NASA)*

House, this successful mission cleared the way for him to announce 20 days later the objective of sending men to the Moon, a response to the orbital flight of Yuri Gagarin on 12 April.

Mercury Redstone-4/spacecraft 11 *Liberty Bell 7* (Virgil Grissom)
Launched: 21 July 1961
Flight duration: 0hr 15min 37sec
Second and final precursor to orbital flight launched on the MR-8 launcher carrying for the first time a capsule equipped with the large observation window lobbied for by the

astronauts early in the development design phase. The spacecraft performed well, reaching an altitude of 118.3 miles (190km), a downrange distance of 302 miles (486km) and a speed of 5,168mph (8,315kph). Grissom's flight was modified to include full manual control on all axes simultaneously. No engineering changes were driven by specifics of the mission but the reason for the unexpected jettisoning of the explosive side hatch after splashdown was never resolved. The spacecraft sank and was lost, Grissom being recovered to the USS *Randolph*.

Mercury Atlas-4/spacecraft 8A (unmanned)
Launched: 13 September 1961
Flight duration: 1hr 49min 20sec
This was the first successful launch, orbiting and recovery of a Mercury spacecraft and had a man been on board it would have been a success. The capsule was the refurbished spacecraft previously launched on the MA-3 flight, hence the 'A' suffix, and it carried a human simulator. The spacecraft was a very complete example of a definitive capsule although the landing bag was not installed and it had the small viewing window and not the large observation window. The side hatch did not have an explosive-release feature and the instrument panel had a unique configuration.

The Atlas 88-D placed the spacecraft in an orbit of 142 miles x 99 miles (228km x 159km) on the second attempt at an orbital flight with a device simulating the metabolic conditions that would be imposed on a pilot. Anomalies occurred in the three spacecraft systems, which could have jeopardised the success of the mission. These included an inverter failure during powered ascent, a leak in the oxygen supply and a range of separate but distinct malfunctions in the spacecraft attitude control system.

The faulty inverter failed due to excessive vibration during powered flight off the launch pad but it was switched out of the circuit and a standby inverter switched on in its place. An astronaut would have been able to do this instantly but if ground command had been unable to send a corrective command to switch inverters it would have resulted in an abort.

The leak in the oxygen system resulted from relaxation of spring force, which allowed partial opening of the emergency-rate control handle. Because the flight was planned for only one orbit sufficient oxygen remained to accommodate the leak rate and to have provided the 'pilot' with the required amount prior to splashdown. On a longer mission the leak rate would have required correction and had an astronaut been aboard pressure could have been exerted on the emergency-rate handle to correct the problem. In fact, onboard film coverage of the instrument panel showed that the oxygen supply emergency light had blinked on, which would instantly have attracted a response from an astronaut. Vibration also dislodged the rate handle, allowing a valve to crack open; this resulted in the development of a new latching mechanism to prevent it jumping out of detent.

The malfunctions in the control system concerned the horizon scanner, wiring in the amplifier-calibrator and the two 1lb (4.448N) thrusters. The horizon scanner displayed random error signals that were sent to the attitude gyros because of so-called cold cloud effects, and a continuous 'ignore' signal from the roll scanner during most of the dark side transit of the orbit occurred because of a short circuit in the capacitor. Ignore signals were activated to stop improper slaving of the gyros when the scanner became saturated with incident light. The random error signals were not of sufficient magnitude to warrant further attention.

The opening of the circuit from the pitch-rate gyro to the amplifier calibrator was caused by a loose wire and resulted in erratic motion of the spacecraft and excessive fuel usage in an attempt to null out the dispersions from planned attitude angles. The thruster failures occurred in the roll-left and yaw-right units and the loss of these caused the spacecraft to exceed normal attitude limits in the ASCS mode that activated the high-thrust units; it was the employment of these that caused the excessive fuel consumption. While sufficient fuel remained for the one-pass mission, had this been a standard manned flight of three orbits there would have not been sufficient fuel remaining for a fully automated attitude control mode. There was also some performance degradation in the

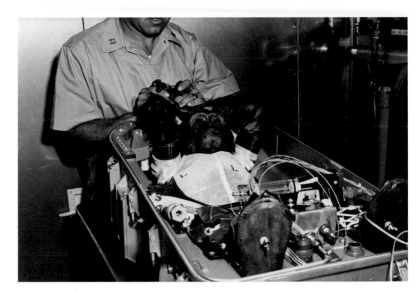

automatic attitude control system, attributable to contamination of metering orifices.

Had a pilot been on board he would have fulfilled a three-orbit mission by switching to the manual proportional or fly-by-wire control mode, thus conserving propellant. With the use of manual proportional mode he could also have prevented the attitude excursions during retrofire that caused the spacecraft to land 176 miles (283km) east of Bermuda, 34 miles (55km) off target. Recovery was by the destroyer USS *Decatur*.

MA-5/spacecraft 9 (chimpanzee 'Enos')

Launched: 29 November 1961
Flight duration: 3hr 20min 59sec
The mission was planned as a three-orbit flight carrying chimpanzee 'Enos' on a full simulation of a mission profile for the initial manned orbital mission. Additional objectives were to evaluate the global Mercury tracking network and wring out any system problems prior to releasing the Mercury Atlas combination to carry an astronaut into orbit. Because of this it was the last of the qualification capsules and carried the large observation window, landing impact bag, a positive lock on the emergency oxygen rate handle, an explosive-type side hatch, new provision for cooling the inverters and modified rate gyros to improve accuracy in a vacuum condition.

The Atlas had a new lightweight telemetry system and a redundant path for sustainer engine cut-off signal so that the ground could send a manual command to prevent an overrun risking

the spacecraft being boosted to a higher orbit than planned. This modification was applied to all Mercury Atlas launch vehicles. This was a major test too of the global tracking network, 17 of the 18 stations being ready to receive telemetry as the spacecraft came in sight.

Launched by Atlas 93-D, it was placed in an orbit of 147 miles x 99 miles (236km x 159km) but the flight was terminated after two passes because of a failure in the control system. Increased fuel consumption caused by a metal chip in a metering device in a low-thrust roll unit caused continuous excursions off attitude to the level where the flight had to be brought down early.

Other problems included a signal to indicate the clock was 18 seconds fast, corrected by a command sent from Cape Canaveral at the end of the first pass; elevated inverter temperatures which had been seen on previous flights but which never reached dangerous levels; and a problem with the environmental control system which produced high cabin temperatures. Enos too got frustrated when one of the three levers he was trained to pull in a specified sequence malfunctioned and gave him a mild electric shock even when operated correctly. The chimpanzee's temperature too increased when water in the felt evaporator pad turned to ice.

The decision to bring the spacecraft back to Earth was taken so that sufficient fuel would remain for the control of the spacecraft's attitude during re-entry. Recovery was carried out by USS

Stormes at the planned landing point and the capsule was retrieved 1hr 15min after splashdown. When the external lanyard was pulled to fire the explosive entry hatch the propagated shock cracked the main observation window.

MA-6/spacecraft 13 *Friendship 7 (John Glenn)*
Launched: 20 February 1962
Launch weight: 4,265.26lb (1,934.72kg)
Orbit weight: 2,986.78lb (1,354.8kg)
Flight duration: 4hr 55min 23sec

The first US manned orbital space flight began with the launch of Atlas 109-D with a spacecraft configured much as the main body of Chapter 3 describes. It was placed in an orbit of 162 miles x 100 miles (260km x 160km) and completed the full three orbits planned. The flight was designed as a demonstration that humans could fly in space, albeit having been launched after two Russians (Yuri Gagarin and Gherman Titov) had already orbited the Earth a total of 18 times the previous year.

Only minor irritations occurred before launch although the cumulative duration of 2hr 17min was frustrating, launch occurring at 9:47am. A minor problem in attitude stabilisation occurred when there was a brief pause in starting the five-second rate-damping cue which induced a significant roll error, quickly brought back into proper alignment. Turnaround fuel expended was 5.4lb (2.45kg) of the 60.4lb (27.4kg) loaded before flight.

Early into the second orbit Glenn executed a planned yaw manoeuvre that saw his capsule turn around 180° to a forward-facing attitude. Some concern was expressed over incorrect readings on the attitude displays compared with his visual cues. However, two of the low thrust units failed again which required the astronaut to take over manual control for much of the flight. Toward the end of the second orbit the thruster problem was picked up by the tracking station in Guaymas, Mexico, but Glenn elected to set the attitude control in fly-by-wire rather than manual proportional because that appeared to use less fuel. Then the system resolved itself and he switched back to ASCS but very quickly thereafter the thruster totally failed and the remainder of the mission was flown manually.

As Glenn completed his second orbit,

flight controller William Saunders reported a Segment-51 signal that indicated the heat shield and landing bag were no longer locked in position. If true only the retro-package restraint straps were holding the shield and bag in place. The astronaut was not informed of this anomaly despite repeated calls at each tracking station he passed for him to leave the landing bag deploy switch in the 'OFF' position.

The obvious solution followed in Mission Control was to leave the retro-package on after retrofire so that the straps would keep it tight against the forebody securing clamp, with the proviso that all three retro-rockets fired as planned. If one rocket failed to fire the heat of re-entry would cause it to explode, rather than burn, potentially damaging the shield itself. An opposing solution, argued by the data analysts, was to release the retro-package as planned, thus avoiding collateral damage from a strapped-up retro-package being burned away as the spacecraft slithered through the atmosphere.

In contact through Hawaii on the last orbit Glenn was asked to cycle the landing bag deploy switch to the 'AUTO' position, which he did, noting that its attendant light did not come on, indicating the heat shield was securely locked in place. Despite this arguing a case for proceeding with the normal mission sequence and jettisoning the retro-package, after consulting with Max Faget it was decided to leave the package on, Glenn manually retracting the periscope and activating the 0.05g sequence by going to manual override. All three retro-rockets fired and Glenn manually resettled the capsule to the -14° pitch-down attitude for entry interface.

There was no measurable effect from the pack remaining attached until it broke free amid flaming chunks of debris hurtling past Glenn's observation window, but with manual fuel supply running low at 15lb (6.8kg) the pilot switched to fly-by-wire and soon after peak g-force the capsule began oscillating well past the 10° point considered safe. Switching to the auxiliary damping system, Glenn nulled some of the attitude excursions in yaw and roll but the ASCS fuel ran out at 11 seconds prior to drogue deploy and the RSCS supply with 55 seconds to go. As the oscillations rapidly increased Glenn opted to deploy the drogue manually, and much earlier than the planned

21,000ft (6,400m), in an attempt to stabilise the capsule and prevent it flipping over apex-forward, but a second before he did that the drogue popped out at a height of 28,000ft (8,534m) and stabilised the falling capsule.

Post-flight, modifications were made to relocate thruster metering orifices and add a new platinum screen material instead of the Dutch-weave fabric, together with more conservative switch rigging and wiring procedures for future capsules. It was also decided that on future flights the astronaut would manually deploy the drogue at the proper altitude rather than rely on the barometric switches; had the drogue deployed much earlier on MA-6 it could have been torn badly or have seriously compromised the release of the main parachute, with fatal consequences.

ABOVE John Glenn shot this image through the window using an Anso Autoset, in reality one of the then-new Minolta Hi-Matic cameras. *(NASA)*

BELOW Glenn is recovered by the destroyer USS *Noa*. The ship later served in Vietnam, was eventually sold to Spain and was scrapped in 1991. *(NASA)*

MA-7/spacecraft 18 *Aurora 7* (Malcolm Scott Carpenter)

Launched: 24 May 1962
Launch weight: 4,244.09lb (1,925.12kg)
Orbit weight: 2,974.56lb (1,349.04kg)
Flight duration: 4hr 56min 05sec

Planned for a three-orbit mission, capsule 18 was similar in all respects to capsule 13 and was launched by Atlas 107-D at 7:45am to an orbit of 167 miles x 100 miles (269km x 160km). The mission's objectives were tweaked so as to allow the pilot greater manual authority over his capsule, with a mixture of engineering and science including planned attitude excursions to view day and night horizons, landmarks and various orientation tests including observation

of the effects on the astronaut of flying upside down, head towards Earth.

This adjustment in programme objectives coincided with the establishment of a committee to explore how scientific experiments could infiltrate the programme without compromising the engineering objectives. One planned experiment was the release of a tethered balloon, 30in (76cm) in diameter housed in the antenna canister along with a gas expansion bottle. SOFAR bombs and radar chaff were removed for the MA-7 flight along with knee and chest straps. The red filter in the window and the heavy Earth observation globe were also taken out, superfluous in the wake of flight experience. Pockets were added to Carpenter's suit for pencils and a handkerchief and the water-wing life vest was installed on the chest below the circular mirror. In the suit circuit the constant-bleed orifice was removed and the landing bag limit switches were rewired in an attempt to prevent faulty telemetry signals in the event one switch indicated a failure.

Carpenter carried different food to Glenn's squeeze-tube options, trying out snacks from the Pillsbury Company and 'bonbons' from Nestlé, processed into bite-sized cubes. Attempting to carry out a packed flight plan, added to by the recommendations of the science group about various tests in orbit, the pilot was unable to devote sufficient time or attention to several systems problems that emerged. The pitch horizon scanner malfunctioned resulting in an excessive signal bias that was not constant in magnitude, causing large errors in pitch attitude while in automatic mode. This scanner problem was not apparent to Carpenter or to ground controllers until quite late in the mission. It made the indication in pitch attitude erroneous, because the indicators on the panel were driven by the output of the attitude gyros and were normally used in precise manual control of the spacecraft.

Carpenter was forced to cross-reference the instrument readings with his view out the window and here he made several errors in the proper energy management of the attitude control system, burning off much more thruster fuel than planned. By the end of the second orbit he had gone through 58% of the manual and 55% of the automatic fuel quantities. To

conserve fuel he went to drifting flight whereby the pointing attitude of the capsule wandered around at will and under the minuscule micro-gravity effects at that altitude.

However, spacecraft systems performed well until late on the third orbit when the true pitch versus indicated attitude were at variance, and because this was noted very close to retrofire there was insufficient time to take corrective action. In fact Carpenter had inadvertently left the manual system switched on when shifting to the fly-by-wire mode. This consumed fuel at a prodigious rate and required Carpenter to manually fire the retro-rockets, but both the attitude of the spacecraft and the ignition time of retrofire was off by three seconds, causing the spacecraft to go long during the entry glide slope. The capsule ran out of fuel during re-entry, which compounded errors that put the spacecraft on the water 288 miles (463km) downrange of the planned landing spot. The astronaut was retrieved 2hr 59min after splashdown and airlifted to the USS *Intrepid*.

MA-8/spacecraft 16 *Sigma 7* (Walter M. Schirra)

Launched: 3 October 1962
Launch weight: 4,324.55lb (1,961.6kg)
Orbit weight: 3,028.89lb (1,373.90kg)
Flight duration: 9hr 13min 11sec

Launched by Atlas 113-D, MA-8 was planned as a six-orbit mission, lift-off occurring at 7:15am to an orbit of 176 miles x 100 miles (283km x 160km). Changes made to the capsule included weight-saving modifications

and improvements to the reliability of components as well as modifications for the longer flight, identified below. Unique also was the elimination of a hold-down delay on the launch pad, lift-off occurring at engine start.

Planning for the extended-duration flight began right after MA-6 where it was seen as a logical step toward the anticipated 22-orbit flight of MA-9. Although the early target for long duration had been a one-day flight all along, in the interests of getting it into space the capsule had been qualified for a four-orbit flight at most. MA-8 was as much a systems qualification flight as it was a development mission. MA-8 duration planning focused on experience to date. The previous two flights had consumed 7,080Wh of

RIGHT It fell to Walter M. Schirra to carry out the final six-orbit shakedown flight of the Mercury spacecraft, flying a near-perfect mission-management flight on 3 October 1962 in spacecraft 16 (*Sigma 7*). *(NASA)*

to free drift, thus conserving electrical power, and by switching telemetry transmitter and radar beacon to ground command the margin could be raised to 15%. The quantity of oxygen required was 4.4lb (2kg) but with contingency reserves for emergencies factored in mission planners needed a total 8.6lb (3.9kg). This could be achieved if the existing design leak rate limit of 61in³ (1,000cm³) could be stiffened up to 36.6in³ (600cm³). Before the flight spacecraft No 16 checked out with a leak rate of 28in³ (460cm³). The additional carbon dioxide exhaled could be effectively scrubbed by increasing the quantity of lithium hydroxide in the existing canister from 4.6lb (2.1kg) to 5.4lb (2.4kg).

battery life from an installed supply of 13,500Wh and a seven-orbit flight was calculated to require 11,190Wh.

With the need for a 10% margin to accommodate contingencies, it seemed prudent to schedule the next flight for six orbits. But, by allowing significant proportions of the flight

The quantity of attitude control fuel required for a six-orbit flight required careful planning of capsule attitude orientation, balancing free drifting flight and thruster control, and balancing automatic versus manual modes. One lesson from MA-7 introduced a control-mode selector lever so that the high-thrust units would only be used for fast attitude correction manoeuvres and would not be used for normal changes in pitch, roll or yaw. Through judicious flight planning mission managers calculated that the automatic system would expend 23lb (10.4kg), leaving 12lb (5.4kg) in reserve, and that the manual system would consume 18lb (8.2kg), leaving 15lb (6.8kg). Also, an additional 16lb (7.3kg) of cooling water was added to the spacecraft.

It had originally been planned to remove the telescope for this flight but as one of the key engineering tests was whether man in the loop could do better than automated systems at conserving attitude control fuel it was felt advisable to carry it. Schirra would use both the periscope and the large observation window to verify attitude, particularly during pre-retrograde changes.

Everything pointed positively to a seven-orbit mission but, when recovery contingency constraints were factored in, it was discovered that the migration of the orbit with respect to the rotating Earth – turning in an anticlockwise motion as viewed from the North Pole – imposed very significant increases in the recovery forces required. By moving to a six-orbit mission that problem was avoided.

One experiment that had been installed was

BELOW Cumulus clouds fill this image shot by Schirra during his MA-8 flight. *(NASA)*

in response to a high-altitude nuclear test on 9 July 1962. Detonated at an altitude of 250 miles (400km), the 1.4-megaton explosion created a new radiation layer, closer to the atmosphere than the natural Van Allen belts. While speculation as to the danger posed to tissue argued each way, a radiation dosimeter was installed on the hatch to determine the increase in radiation experienced by the astronaut.

Following an impeccable lift-off the Atlas began an accelerating roll which came to within 20% of the limits that would have triggered an abort, and then, mysteriously at the time, smoothed out to a normal flight, but an overburn sent it slightly higher at apogee than any other Mercury spacecraft. Arriving in orbit 5min 15sec after lift-off, the capsule separated two seconds later and Schirra began a committed and concerted examination of optional spacecraft control modes, using carefully mapped sequences and some innovative switching methods of his own that saved fuel and resulted in 78% of both manual and automatic fuel quantities remaining. Around 53% remained at auto-fuel jettison after re-entry.

The spacecraft performed well, the actions of the astronaut were impeccable and satisfied engineers and managers that the Mercury spacecraft was capable of a full day in orbit. No failures were detected that could have imperilled the mission but a partial blockage of the coolant control valve for the suit environmental circuit was corrected by careful manipulation of the control. If no means had been available for controlling the temperature in this way the mission would have been terminated.

MA-9/spacecraft 20 *Faith 7* (Gordon Cooper)
Launched: 15 May 1963
Launch weight: 4,330.82lb (1,964.4kg)
Orbit weight: 3,033.35lb (1,375.92kg)
Flight duration: 34hr 19min 49sec
The Manned-One-Day-Mission (MODM), as it became known, was launched by Atlas 130-D at 8:04am to a 166 miles x 100 miles (267km x 160km) orbit. The optimum flight plan envisaged a 22-orbit mission due to orbit phasing and the abort contingency landing points.

Internally, MA-8 marked the end of Project Mercury, the final flight being officially designated

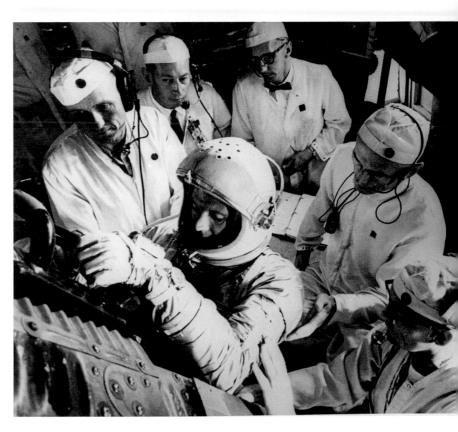

ABOVE Helped by pad technician Guenter Wendt (third from left), Gordon Cooper slips into spacecraft 20 (*Faith 7*) for launch on the longest flight in the Mercury programme. *(NASA)*

LEFT Atlas 130-D lifts off on 15 May 1963, the one-day mission sought by the Air Force and then by NASA to test how well humans could conduct activities in space. Launch Complex 14 had seen its first launch on 11 June 1957 and was last used on 11 November 1996. *(NASA)*

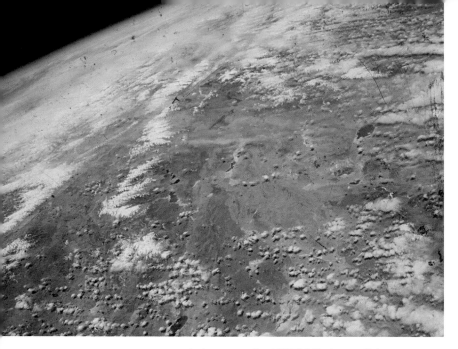

ABOVE **A view of Tibet as seen through the window of** *Faith 7.* *(NASA)*

BELOW **The carrier USS** *Kearsarge,* **recovery ship for the MA-9 flight, proudly displaying its vital role in the Mercury programme.** *(NASA)*

in-house as the MODM Project, utilising a Mercury spacecraft. The redesignation never caught on publicly, the media continuing to report it as one of the continuing missions under Project Mercury; the change was a mere semantic.

In all there were 215 changes to spacecraft 20 including removal of the Rate Stabilization Control System (RSCS), replacement of the pitch attitude gyro with a caged gyro slaved to -34° retrofire attitude, improvements to the 1lb (0.45kg) and 6lb (2.7kg) thrusters, nitrogen tank pressure in the ASCS system increased to 2,800lb/in² (19,306kPa), an additional 15lb (6.8kg) water tank added in parallel with the automatic one, a manually operated fuel transfer valve between automatic and manual fuel systems and inverters cooled by heat sinks rather than coolant circuit and heat exchanger.

The CO_2 absorption capability was increased by an additional 0.8lb (0.36kg) of lithium hydroxide, an additional 4lb (1.8kg) bottle of oxygen was added in parallel to the primary bottle, an additional 9lb (4.1kg) bottle of coolant water in parallel with the existing 39lb (17.7kg) tank was installed, multiple discriminatory changes to the space suit were made including a urine-transfer fitting, a 'work table' was added in front of the instrument panel and an isometric exercise device similar to that which had been carried on MA-6 was provided in the capsule.

For power production, a redundant 3V supply was added for instrumentation reference which could be activated by the astronaut in the event of a failure to the primary system. The eight-day correlation clock (MA-8) was removed and the satellite clock was powered by the 24V DC bus rather than the isolated bus. More importantly, increased power was provided in the form of five 3,000Wh and one 1,500Wh batteries in place of the three 3,500Wh and three 1,500Wh batteries fitted to MA-8.

Several changes were made to the communications system and a slow-scan television system was added for real-time observation of the pilot and the spacecraft environment. The TV transmitter was also available as a backup to the telemetry transmitter. In flight the TV system proved satisfactory and all transmissions were received by ground stations but the pictures were of poor quality due to inadequate interior lighting and observation of Earth features outside the spacecraft were washed out by too much sunlight.

Science and observation equipment added included a tethered balloon similar to that flown during the MA-7 mission, packaged in the antenna canister, a self-contained flashing beacon in the retro-package, two Geiger counters, a dosimeter, four film badges and a Schaeffer radiation package to determine radiation exposure during the flight. Spacecraft 20 also carried a 70mm Hasselblad 500-C modified for taking general colour photographs, infrared weather pictures and horizon definition photographs during the MA-9 mission. Cooper also had use of a special 35mm camera for photographing zodiacal light and the airglow layer.

Cooper was able to adjust around several minor but irritating technical problems including the drinking water valve, the condensate transfer system and an indication that the 0.05g light had come on unexpectedly. Checks indicated that the amplifier-calibrator was in the 0.05g configuration and that the ASCS could only be used during re-entry. Then, at an elapsed time of 33hr 7min, a short circuit occurred in a busbar serving the 250V main inverter which took power from the ASCS, and a manual re-entry was performed flawlessly. Splashdown occurred a mere 4.6 miles (7.4km) from the designated location and the astronaut was lifted from the waters of the Pacific Ocean, inside his capsule, 41 minutes later and taken to the USS *Kearsarge*.

Early in the flight Cooper had a problem with the suit temperature and had to make several adjustments but the problems were minor and correctable with the presence of a man in the loop – which was the fundamental objective of the entire programme, to determine just how effective a pilot could be in controlling a spacecraft and managing systems and subsystems which, for the most part, were designed for autonomous operation.

MA-10

On 23 October 1961 tacit approval was given for four one-day missions to follow the initial fulfilment of the basic Mercury programme and expand on the spacecraft's capabilities. Spacecraft 12, 15, 17 and 20 had been set aside for these extended-duration flights with a final contract negotiation with McDonnell settled in September 1962 for appropriate modifications. MA-9 was the first of those, known as the MODM mission programme. MA-10 was to have been a three-day flight piloted by Alan Shepard.

However, there had already been calls for the Mercury programme to close after the successful flight of MA-8 and as early as April 1962 NASA headquarters had notified its intention to bring an end to Mercury after MA-8 and MA-9. The agency was hard at work on the two-man Gemini derivative of Mercury, while still in the final planning stages for how to fly Apollo to the Moon. But the final decision on MODM was still pending when Cooper flew his 22-orbit mission.

Based on the several systems problems that emerged during Cooper's MA-9 mission, NASA administrator James W. Webb met with senior managers on 6–7 June to determine the fate of MA-10. Aware that his mission might be cancelled, Shepard lobbied for it to be flown, unsuccessfully appealing to President Kennedy. But Webb decided on 12 June 1963 that the programme would shut down, switching resources to the Gemini programme; there was little purpose in continuing a programme already hard up against the buffers in terms of capability and mission-extension. Resources were desperately needed elsewhere, but these last two Mercury flights had been invaluable in plugging gaps in engineering and physiological knowledge useful in the run-up to Gemini.

Aftermath

Seven men had been picked to fly the Mercury spacecraft. Donald Slayton had been selected for MA-7 but a medical examination discovered that he had an erratic heart rate and in September 1962 he was grounded. Had these additional MODM flights been authorised, Virgil Grissom would have been assigned to MA-11. There was one further opportunity for Shepard and he was promised command of the first Gemini mission with rookie astronaut Thomas Stafford. But late in 1963 Shepard went down with Ménière's disease, a build-up of pressure in the inner ear, which took him off the roster. In his place Grissom was given the command pilot seat on the first manned Gemini mission (GT-III), along with another rookie, John Young.

After surgery Alan Shepard was returned to flight status in May 1969 and went on to command Apollo 14, the third Moon landing, in January 1971. By cutting down alcohol consumption, stopping smoking and taking a large range of vitamins and supplements, 'Deke' Slayton's fibrillating heart returned to normal and he too returned to flight status. In July 1975 he flew on the joint Apollo-Soyuz docking flight where astronauts met cosmonauts and shook hands in space, ending a divide epitomised in the minds of many by the race for space and America's first manned space programme.

The Mercury spacecraft was seen as a useful tool for applications outside the specific objectives of manned space flight and while contractors are naturally always keen to point out derivatives and variants applicable to other objectives, which call for extended production lines, a wide range of

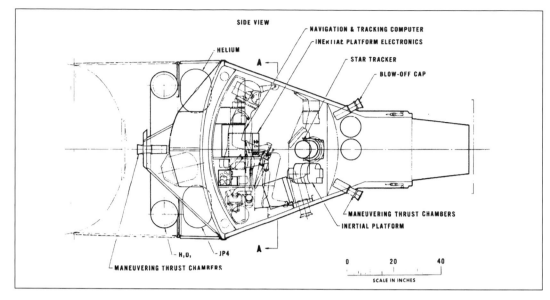

proposals were being considered by NASA itself, even when the roadmap for future human space flight had been set down by President Kennedy in May 1961 when he declared the Moon-landing goal. Some of these were proposed as a bid for specific field centres to attract wider control over existing programmes.

On 1 November 1961 the Space Task Group had been renamed the Manned Spacecraft Center and after the completion of the operational phase of Project Mercury the team would move to a new facility outside Houston, Texas, which became famous as the place where the voice of Mission Control spoke to the world about dramatic events in space and on the Moon. Eager to make its presence felt within NASA as a new and rising star, which very quickly would come to dominate the space

agency's budget, proposals began to emerge from MSC for new uses for Mercury, one of which addressed concerns about the frequency and distribution of micrometeoroid particles in space that posed a hazard to the safety of manned vehicles.

To quantify the size of the problem and to qualify decisions on material thickness in the design of the Apollo spacecraft pressure shell, engineers wanted to expose a sheet of aluminium, held coiled in a drum until deployment in orbit, to measure their effects. Mercury, Gemini and Apollo used pressure-cabin skin thickness of 0.17in (0.43cm) but to gain an order of magnitude in scaling down the thickness of the detector sheet to 0.016in (0.04cm) they would reduce by three orders of magnitude the mass of the penetrating

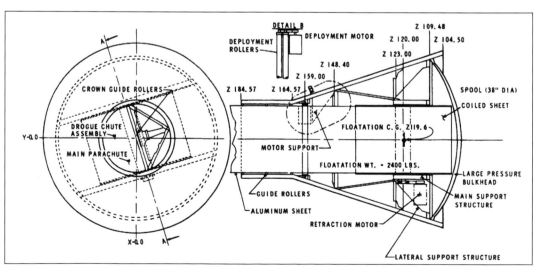

meteoroid from 10^{-4} to 10^{-7}gm. That would effectively provide a quantitative base from which to extrapolate penetration probabilities.

NASA-MSC proposed adapting a Mercury spacecraft to carry a drum of aluminium-coated Mylar formed so as to be rolled flat when coiled but to spring open to a tube when deployed. The deployed tube would have been 1,846ft (563m) in length with a width of 26in (66cm) when flat and 8in (20cm) in diameter when tubular on deployment, providing a surface area of 3,500ft² (325m²). To prevent buckling during retraction at the end of the mission it would have to be rolled back up at a maximum rate of 1in/sec (2.54cm/sec) initially, increasing to 3.5in/sec (8.9cm) toward the completion of the process, requiring just over 5.5 hours, or 3.5 orbits.

With a thickness of 0.0003in (0.00076cm) the aluminium foil covering would act as a recuperable capacitance gauge when connected to the terminals of the on-board batteries, a process which had never been used in space to that date. The unmanned modified Mercury would conduct a 14-day mission after which it would return to Earth with the recoiled drum for analysis. At launch the spacecraft would weigh 2,983lb (1,353kg), in orbit it would weigh 2,797lb (1,269kg) and after splashdown it would weigh 2,398lb (1,088kg). The coiled drum and associated experiment equipment would comprise 1,139lb (516kg) of those totals.

The concept was a sound idea but there were too many technology unknowns to warrant development. Concurrently, the Marshall Space Flight Center was developing the Pegasus micrometeoroid payload which would be launched into space by, and remain attached to, the cryogenic S-IV second stage of the Saturn I launch vehicle. Three such configurations would be launched successfully during 1965.

Triggered largely as a result of Gordon Cooper's 22-orbit flight in May 1963, the Air Force revived its interest in Mercury as a possible photoreconnaissance platform, at a time when the National Reconnaissance Office was about to launch the first of its very-high-resolution Gambit spy satellites. Before the end of that year both the NRO and the Central Intelligence Agency were working toward a broad area search system that would eventually become the Hexagon programme and the

KH-9. Some specialists advising the Air Force persuaded them that an adapted Mercury spacecraft could carry a very-high-resolution camera for returning film to Earth but its limited capabilities in both weight and pointing angles doomed the proposal.

There was one last push for further applications of the Mercury spacecraft. In 1964 the Langley Research Center conducted a study to use unmanned Mercury spacecraft as orbiting astronomical observatories. There was in existence at the time an official NASA Orbiting Astronomical Observatory programme run by Goddard Space Flight Center, these being launched from 1966. But Langley promoted the use of a recoverable spacecraft for returning to Earth with film exposed in space. The existing OAO satellites being built by Grumman were non-recoverable and transmitted data back to Earth via radio links.

The Mercury OAO would have carried a cassegrain reflecting telescope with a 30in (76cm) aperture achieving a gain of up to ten over telemetered systems using the fine stabilisation control holding drift to less than 0.08arc-sec during exposure. The adaptation would have required complete removal of the internal pressurised section and the recovery section to accommodate the optics, the latter replaced with a toroidal structure surrounding the forward end of the afterbody. The Mercury OAO would have weighed 4,740lb (2,150kg) including 1,345lb (610kg) for the telescope and would have required an Atlas-Agena B launch vehicle. The concept was never pursued.

BELOW One proposal from the Langley Research Laboratory would have had McDonnell modify the Mercury spacecraft as an orbiting astronomical observatory, but this failed to make headway at NASA headquarters. *(McDonnell)*

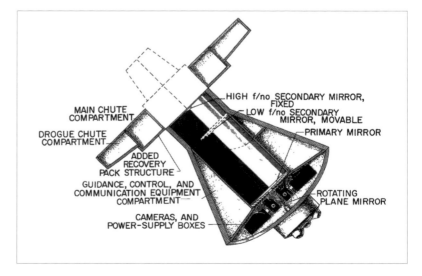

Chapter Five

The launch vehicles (1959–63)

The first rockets to lift US astronauts into space were modified Redstone and Atlas ballistic missiles, but the one which would test the capsules, and the escape system, was NASA's first dedicated rocket, Little Joe. Together, all three would qualify America's first manned spacecraft for humans to safely ride beyond the atmosphere.

OPPOSITE Carrying the monkey Miss Sam, Little Joe-1B was launched on an escape system test on 21 January 1960. *(NASA)*

ABOVE **The flight of Little Joe-1, the first launch in the Mercury programme, took place on 21 August 1959.** *(NASA)*

Little Joe

The idea for this clustered set of solid-propellant rockets arose from a concept worked out by Max Faget and Paul Purser in January 1958 for propelling a manned capsule on a ballistic trajectory in a plan called High-Ride. Similar to the Army's Project Adam, it soon lost favour as the primary means of getting a man into space. However, using combinations of Castor and Pollux main motors and Recruit auxiliary rockets it was ideal as a cheap and effective way of testing certain aspects of the flight profile of a Mercury spacecraft, including maximum aerodynamic

pressure, launch escape system and recovery techniques.

The name Little Joe was coined when early drawings showing four exhaust holes looked like a double deuce on dice, referred to in the jargon of the American game of craps as a 'Little Joe'. The names Castor and Pollux signified two variations of the MGM-29 Sergeant tactical battlefield missile, an Army rocket developed by the Jet Propulsion Laboratory (JPL). (Castor and Pollux are the two stars in the constellation Gemini.) Each Atlas cost $2.5 million and a Redstone $1 million per flight; the $200,000 spent on each Little Joe was a good return on the minimal cost of development and provided a relatively cheap means of qualifying the launch escape system. When NASA put out a request for manufacturing Little Joe, 12 airframe companies responded and North American Aviation won a contract on 29 December 1958.

With a height of 8.29m (27.2ft) and a diameter of 2.03m (6.7ft) one configuration of Little Joe would consist of a cluster of four XM-33 Castor and four XM-191 Recruit solid-propellant rocket motors. With a manned capsule and its launch escape tower attached, the configuration had a height of 15.16m (49.75ft) and would weigh about 19,700kg (43,440lb). Four large splayed tail fins provided a maximum span of 6.5m (21.3ft), installed to ensure stability around Mach 6. This had an adverse effect in that it caused the vehicle to weathercock into wind at low speeds, which

RIGHT **The Little Joe rocket was assembled in a variety of configurations to suit specific mission requirements. This arrangement was for the LJ-2 flight on 4 December 1959.** *(NASA)*

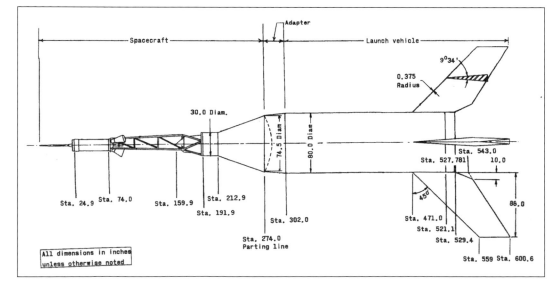

was resolved by configuring the solids so as to ensure high acceleration at lift-off.

In a typical launch with an eight-cluster application, the four Recruit and two Castor rockets ignited to produce a thrust of 1,156kN (260,000lb), the second pair of Castors igniting at 23 seconds followed by burnout of the other two and the four Recruits seven seconds later. The remaining two Castors continued to burn until an elapsed time of 58 seconds into flight. Little Joe could lift a maximum 1,788kg (3,942lb) with typical maximum altitude for the capsule being 85km (53 miles).

Various configurations of solid propellant rockets could be clustered in four basic configurations of Little Joe. These were: Type I: 2 Castor x 4 Recruit; Type II: 4 Castor x 4 Recruit; Type III: 2 Pollux x 4 Recruit; Type IV: 4 Pollux x 4 Recruit. The defining element of selection for the different types was the required performance for specific tests. Castor and Pollux had the same dimensions but the latter had greater thrust.

Mercury Redstone

America's first manned space launcher was developed out of a ballistic missile programme that had seen previous derivatives launch satellites for the Army and for NASA. The redesign of the basic Redstone missile required enhanced performance, which was obtained through increased burn time, a simplified control system and the addition of an in-flight abort-sensing system, these collectively comprising a man-rating programme that saw more than 800 changes to the basic rocket. The mission objective was to flight-test unmanned Mercury spacecraft and to launch astronauts on suborbital flights to space and back.

The Mercury Redstone programme began with a requirement for eight flight tests, ordered on 3 November 1958, but only six were flown due to the success of the programme and the need to move orbital flights to the Atlas (which see). It had been intended to fly only the first two unmanned but failures and technical problems delayed that until the fourth flight. The first static test of the MR-1 booster took place on 7 January 1959.

The first launch of a Redstone missile had

LEFT The MR-2 launch on 31 January 1961 carried the chimpanzee Ham on a ballistic trajectory. (NASA)

BELOW The flight configuration of the Mercury Redstone. The rocket had already achieved fame by being a heritage development of the ballistic missile that also placed the first American satellite in orbit three years to the day before the MR-2 flight. (NASA)

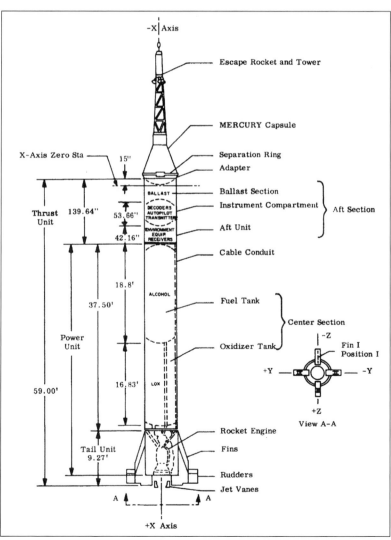

occurred on 20 August 1953 in a development programme that culminated in a Block II version providing the Army with a missile that had a range of 200 miles (322km). By January 1959 there were two versions, the advanced Block II version utilising an improved A-7 engine with liquid oxygen and alcohol as propellants and the Jupiter-C with the A-5 engine replacing the alcohol with the more toxic Hydine consisting of 60% unsymmetrical dimethyl hydrazine and 40% diethylene triamine. It was this extended-performance four-stage version that had been used to launch the first US satellite, Explorer 1, on 31 January 1958.

The Mercury Redstone rocket was based on the Block II A-7 version but with enlarged propellant tanks of the Jupiter-C, which increased the burn time from 123.5sec to 143.5sec. By the first manned launch on 5 May 1961 the 69 Redstone rockets launched to that date had achieved a reliability rate of 81% and the Block II had achieved 11 consecutive successes while Jupiter-C had achieved seven consecutive successes.

Mercury Redstone was less stable in flight than the standard Redstone, especially in the supersonic region, and 687lb (312kg) of ballast was added forward of the instrument compartment to help overcome this. Some changes were also made as a result of the decreased lateral bending frequencies, which were reduced to one-quarter those of a standard Redstone, inciting resonance problems during flight tests with the second bending mode filtered out of the control system to prevent feedback.

The requirements of the ballistic trajectory carried the rocket vertically from its Cape Canaveral launch pad for the first 30 seconds, causing difficulties for range safety in that destruction during this period could bring debris down on the site itself and the ground control centre. The flight path was shaped to a nominal Earth-fixed velocity of 6,500ft/sec

LEFT Unrelated to the Mercury programme, this shot displays the way the erector-trailer was winched from a horizontal position on its trailer bed to a vertical position for placement on the launch structure at the pad. *(USAF)*

(1,980m/sec) with an injection angle of 41.8°, a cut-off altitude of 200,000ft (60,960m) with an acceleration of 6.3g at 2min 23sec.

Operating on ethyl alcohol and liquid oxygen propellants, the A-7 motor delivered a sea-level thrust of 78,000lb (346.944kN) with a turbopump driven by hydrogen peroxide. The pneumatic control system provided gaseous nitrogen to operate the propellant and hydrogen peroxide valves and to pressurise the peroxide tanks. The automatic in-flight abort system was the nation's first attempt at man-rating an existing launch vehicle, driven by the need for exceptionally high standards of quality control as well as the provision of diagnostic and control systems specific to that goal.

The automatic abort system was developed to detect vehicle malfunctions that could compromise the safety of the astronaut, and was configured to shut down the rocket motor and send an abort signal to the spacecraft. This would activate the Mayday relays described in the 'Rocket systems' section in Chapter 3, and the abort sensing systems had to be compatible with the launch vehicle, the capsule interface and the various modes of operation. It was universally accepted that this alone was the greatest single item in the Mercury programme that improved crew safety.

The physical characteristics of the Mercury Redstone configuration gave it a height of 83.38ft (25.41m) from the base of the four equally spaced triangular fins at the bottom of the rocket to the top of the launch escape system. A typical Mercury Redstone flight would have a lift-off mass of about 65,800lb (29,850kg) of which 52,159lb (23,655kg) would be propellant consumed during ascent.

One aspect of the Mercury Redstone project that has received little discussion was the engineering design of a stage recovery system that would return spent stages to Earth for reuse. Several development steps were taken with the idea of packing a recovery parachute

RIGHT Again unrelated to the Mercury missions, this view shows the erector-trailer vertical on the near side of the rocket. Having placed the Atlas in the vertical position the erector-trailer is lowered and driven away by a tractor vehicle. *(USAF)*

in the forward section of the booster casing but the idea was abandoned when there were insufficient funds to carry the concept forward to a flight-test stage.

Mercury Atlas

Selection of Atlas as the Mercury launch vehicle for America's first orbital flights was made at a time when the rocket was still in the development stage and had yet to prove itself safe and reliable.

Atlas had evolved as America's first intercontinental ballistic missile (ICBM) during the early 1950s and the basic rocket was a 1½-stage design, in that it had a pair of vertically stacked propellant tanks providing kerosene and liquid oxygen to a single sustainer motor as well as two booster motors, one each side mounted on a single skirt attached to the thrust structure. The booster

ABOVE The launch tie-down clamps for the Atlas rocket. Here the rocket is absent for a clearer view. *(USAF)*

BELOW Launch Complex 14 at Cape Canaveral on 14 May 1963, the day before the launch of the last Mercury flight. Looking south in the general direction of Port Canaveral are several more Atlas launch pads. *(USAF)*

President Eisenhower. First launched on 23 December 1958, Atlas C had an improved guidance package, weight-saving measures and refinements resulting from the previous test flights.

Atlas D, first launched on 14 April 1959, was the definitive ICBM and this model was used by NASA for development into the Mercury launch vehicle through man-rating of the existing rocket. One of the most significant changes was the addition of a new autopilot gyro package considerably more advanced than that on a standard Atlas D. This was made necessary by the longer payload and the effect this had on the flexible Atlas tank during flight, providing optimum attitude-rate sensing with minimum engine deflections for more efficient performance.

The Atlas had retrorockets that fired after sustainer engine shutdown so that the rocket stage would slow down and not bump into the payload after it separated, which on an Atlas missile would be the re-entry warhead. These were deleted on Mercury Atlas and the positive separation required to prevent re-contact and possible damage was transferred to the three posigrade rockets on the Mercury capsule so that, after separation, they would push the spacecraft forward to incur an increasing separation. To protect the top of the Atlas tank and prevent it from bursting, additional fibreglass shielding was attached to the mating ring and this covered the entire forward dome.

The guidance system too required change because the trajectory would differ significantly from that when it was used to throw a warhead to its target. New antennas were required to ensure maximum signal strength during powered flight. By far the most important addition was the Abort Sensing and Implementation System (ASIS) designed to accommodate malfunctions in the performance of the rocket by activating the abort features, including the spacecraft's launch escape system. Detailed analysis of all previous Atlas flights showed that in most cases where the rocket was destroyed there were sufficient indicators ahead of the failure, which, if properly sensed in flight and reacted to through an emergency escape procedure, meant that the

motors were separated as a single unit a little more than two minutes after lift-off, leaving the sustainer to continue burning for a further three minutes. Designed as an ICBM it was not intended to achieve orbital flight, the velocity required being greater than it had achieved as a ballistic missile throwing a nuclear warhead 5,000 miles (8,045km).

The first Atlas launch attempt on 11 June 1957 was a failure and the first successful flight was achieved on 17 December, the third attempt. But this was with an Atlas A, which only had the two booster engines attached; the triple-motor Atlas B first flew on 19 July 1958. The first Atlas to orbit the Earth was launched on 18 December 1958 carrying a package of instruments and equipment for broadcasting a pre-recorded message from

capsule could be saved and the recovery of the astronaut assured.

Sensors were designed into the ASIS which would read liquid oxygen pressure, differential pressure between the intermediate bulkheads, attitude rates in all three axes, rocket motor injector manifold pressures, sustainer engine hydraulic pressure and the launch vehicle AC power. Dual sensors were designed into each of these sectors and deviation beyond acceptable boundaries would cause the ASIS to pull power from the 28V DC ring being applied to the detection relays. This voltage drop provided a further level of safety in that should the abort system itself fail the loss of power would trigger an immediate abort; in this way, if the ASIS was no longer active it was not considered responsible to continue the flight and it would be terminated.

Atlas contractor General Dynamics Astronautics developed the abort system with the help of the Air Force and with the coordination of NASA. Built into the self-destruct package was a three-second delay after the engine shut-down command triggered by an abort, so that the Mercury spacecraft could depart the ascending stack and outrun an expanding fireball that could ensue. Atlas was guided by radio command in concert with a Burroughs computer at the ground guidance station. But the path in which the signals from the ground computer to the launch vehicle ran, via the range safety command transmitter and on to the airborne receiver and engine

relay control, was not completely redundant. No duplication path existed in the computer calculating the time to shut down the sustainer engine and this factor represented the last single failure mode remaining.

To remove this potential failure, the existing Azusa baseline interferometer supporting the Atlantic Missile Range off Cape Canaveral was introduced. Operating in the C-band it transmitted to a transponder in the Atlas in conjunction with an IBM IP 7090 computer that continuously computed what was referred to as the instantaneous launch-vehicle impact point, or IIP. The modifications to the IP 7090 allowed it to obtain the precise time at which orbital velocity was achieved and this was provided electrically by landline to the Flight Director, who could use this signal as a backup in the event of a failure in the existing MOD III guidance system on the Atlas.

A typical Mercury Atlas configuration had a total height of 94.3ft (28.7m) from the base of the booster engines' skirt to the tip of the aerodynamic spike on top of the launch escape tower. The height of the Atlas launch vehicle to the base of the spacecraft adapter was 65ft (19.8m) and the launch vehicle had a maximum tank diameter of 10ft (3m) and a width of 16ft (4.9m) over the booster skirts. It had a launch mass of about 260,000lb (117,934kg) and produced a lift-off thrust of 367,000lb (1,632.4kN), of which 300,000lb (1,334.4kN) was from the two boosters and the remainder from the single sustainer.

Appendix 1

Worldwide Tracking Network (WWTN)

Development of a ground-based tracking and communications network began shortly after NASA formed and with the advent of the Mercury programme a global network was essential, with the aspiration that contact with an orbiting spacecraft should be made at least every 15 minutes. In February 1959 Langley Research Center set up its Tracking and Ground Instrumentation Unit (TAGIU) aided by the Goddard communications centre at Goddard Space Flight Center, Greenbelt, Maryland. These two units were to share responsibility for setting up the Mercury worldwide network.

Prospective bidders received the preliminary plan for how that was to be achieved during an informal meeting on 2 April, with formal release of the specifications and requirements on 21 May. By the deadline of 22 June the seven submissions were received and the prime contractor was announced on 13 July as Western Electric Company of New York City. Working with WEC would be Bendix Corporation, Burns & Roe Inc, IBM and Bell Telephone Laboratories Inc.

The plan for the WWTN was to establish a main control centre at Cape Canaveral with a backup control centre at Bermuda, two accessory stations to Cape Canaveral on Grand Bahama Island and Grand Turk Island with a central computer complex at Goddard and 14 smaller tracking stations around the world. Locations were based on maximum coverage for a three-orbit mission, the use of existing facilities where possible and the location of foreign stations in 'American-friendly countries'.

By the end of April 1960 all the necessary agreements were in place with the Department of Defense and the foreign countries involved. Six stations would be in the United States, two US Navy ships would be modified for use as tracking stations in the Atlantic and Indian Oceans and ten would be in foreign countries. Eight existing facilities in the US would be expanded or converted, including locations at Cape Canaveral and Eglin, Florida; Kauai, Hawaii; Point Arguello, California; and White Sands, New Mexico.

Outside the US there were facilities at Woomera, Australia, and accessory stations were at Grand Bahama Island and Grand Turk Island. New facilities were built in Corpus Christi, Texas; Bermuda; Grand Canary Island; Kano, Nigeria; Zanzibar; Muchea, Australia; Canton

BELOW Mission Control Center at the Cape Canaveral Air Force Station was used to manage the Mercury flights and the first three Gemini missions, the last of which, in March 1965, was manned. As a result of the decision to significantly expand the human space-flight programme by sending men to the Moon, the Space Task Group became the Manned Spacecraft Center and moved into new facilities built near Houston, Texas, between 1963 and 1965. *(NASA)*

Island; and Guaymas, Mexico. All stations were declared operational on 31 March 1961.

What became the Mercury Control Center (MCC, an acronym eventually referring to Mission Control Center) was an adapted building at Cape Canaveral, and as the command centre it would direct all flight operations, monitor the aeromedical status of the astronauts and spacecraft systems, have decision over ground-activated aborts, command re-entry, keep the astronauts and tracking network appraised of events, coordinate and monitor the flow of communications worldwide, and inform the recovery forces of key events. To facilitate these functions there were to be 14 flight controllers:

ABOVE Inside the **Mission Control Center at Cape Canaveral from where all Mercury flights were controlled after lift-off.** *(NASA)*

- Operations Director – a NASA employee who would supervise operations and make official decisions whether to launch or scrub based on recommendations from the Flight Director.
- Flight Director – NASA employee to supervise all activity within the MCC and make major decisions on the conduct and prosecution of the mission.
- Flight Dynamics Officer – NASA employee advising on launch readiness and monitoring the launch trajectory and orbital insertion.

- Capsule Communicator (CapCom) – NASA astronaut monitoring all aspects of the flight and conducting communication with the astronaut in space.
- Flight Surgeon – a military doctor observing and monitoring the physiological and psychological condition of the astronaut.
- Capsule Environment Monitor – NASA employee observing the spacecraft's ECS during all phases.
- Capsule Systems Monitor – NASA employee

ABOVE Inside the **Mission Control Center at Cape Canaveral from where all Mercury flights were controlled after lift-off.** *(NASA)*

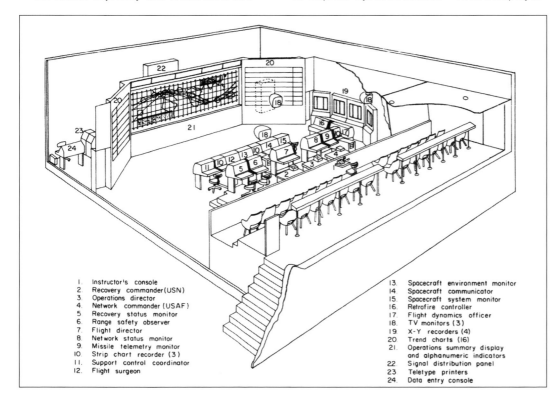

1. Instructor's console
2. Recovery commander (USN)
3. Operations director
4. Network commander (USAF)
5. Recovery status monitor
6. Range safety observer
7. Flight director
8. Network status monitor
9. Missile telemetry monitor
10. Strip chart recorder (3)
11. Support control coordinator
12. Flight surgeon
13. Spacecraft environment monitor
14. Spacecraft communicator
15. Spacecraft system monitor
16. Retrofire controller
17. Flight dynamics officer
18. TV monitors (3)
19. X-Y recorders (4)
20. Trend charts (16)
21. Operations summary display and alphanumeric indicators
22. Signal distribution panel
23. Teletype printers
24. Data entry console

LEFT Mission Control Center console positions together with screens and display boards, a forerunner of its successor at the Manned Spacecraft Center in Houston, famous around the world for managing every human space flight since Gemini IV in June 1965. *(NASA)*

RIGHT Katherine Johnson played a highly significant role in manually verifying the then flaky world of electronics and calculating devices by carrying out complex mathematical calculations to construct space flight trajectories for the Mercury missions. She had been working for NACA since 1953 and was one of several African-American women who were employed for this exacting duty. With clear leadership skills, she had been extracted from the computing pool to engage with other engineers. 'The early trajectory was a parabola, and it was easy to predict where it would be at any point,' Johnson said. 'Early on, when they said they wanted the capsule to come down at a certain place, they were trying to compute when it should start. I said, "Let me do it. You tell me when you want it and where you want it to land, and I'll do it backwards and tell you when to take off." That was my forte.' After verifying trajectory calculations for several missions, including the first Moon landing in 1969, Katherine retired from NASA in 1986. *(NASA)*

observing all non-ECS systems throughout flight.
- Retrofire Controller – NASA employee advising mission duration, required retrofire parameters, systems condition and launch abort retrofire timing.
- Recovery Status Monitor – Navy officer reporting to Flight Director on status of recovery forces.
- Missile Telemetry Monitor – Convair employee monitoring the Atlas launcher and any emergency situations calling for an abort.
- Network Status Monitor – Defense Department employee monitoring global network status.
- Range Safety Observer – NASA employee observing MCC activity throughout and advising of any criteria at Cape Canaveral regarding range readiness.
- Network Commander – Air Force officer commanding the Mercury Range and ordering response to any malfunctions.
- Recovery Task Force Commander – Navy officer commanding recovery operations.

In addition to the above the MCC had an Operations and Procedures Officer who served as the Flight Director's wingman and was usually a NASA employee managing procedures for each console for a specific flight, ensuring procedures were adhered to thereafter. He also coordinated teletype communications throughout the network.

The station at Bermuda was considered the secondary control point for Mercury since it had

LEFT Christopher Columbus Kraft, Flight Director during the Mercury programme, managed every mission up to Cooper's one-day flight where he shared shifts with British-born Canadian John Hodge. Kraft gathered around him a team of expert flight controllers whose names would be enshrined within the annals of space flight history. The men who managed Alan Shepard's brief hop into space as America's first astronaut also put boots on the Moon, flew 133 successful Shuttle missions and built a 400-ton space station now in Earth orbit. Kraft served as director of Houston's renamed Johnson Space Center from 1972 to retirement in 1982. *(NASA)*

responsibility for aborting the mission as the launch vehicle and spacecraft climbed out of range of the Cape. It had essentially the same layout and technical capabilities as the MCC but retained only the following console positions: Flight Supervisor, Flight Dynamics Officer, CapCom, Flight Surgeon, Capsule Environment Monitor and a Capsule Systems Monitor. It also had a unique post – a Maintenance and Operating Supervisor.

The other stations were relatively small in comparison, with the exception of White Sands and Eglin, which were equipped with voice and telemetry equipment. Stations at Muchea, Kauai, Guaymas and Point Arguello were also equipped with a command capability for emergency situations and usually had an astronaut present for communications with the spacecraft, and all stations in the US – as well as Muchea – had direct voice contact with MCC at Cape Canaveral. Usually the 14 remote stations had four flight controllers including the Flight Supervisor (who could also act as CapCom), a doctor, an engineer trained in Mercury onboard systems and a radio/teletype officer for communications with MCC.

A test of the global tracking network was planned using what was designated a Mercury-Scout configuration. The solid-propellant Scout consisted of four stages, had a length of 70ft (21.3m), a weight of 38,863lb (17,628kg) and a first-stage thrust of 105,665lb (470kN) from

an Aerojet Algol rocket motor. The idea was to orbit a 67.5lb (30.6kg) package of instruments that could be used to test and qualify the various tracking stations set up around the world. The specific rocket selected was a modified Blue Scout II and the launch attempt was made on 31 October 1961, after the two manned suborbital flights but prior to the first manned orbital attempt. Shortly after lift-off the vehicle lost control and began to break apart 28 seconds into the flight, a failure caused by incorrect wiring by a technician who transposed two critical electrical leads. Plans for another attempt were cancelled as Mercury Atlas orbital flights were now taking place to do the verification job.

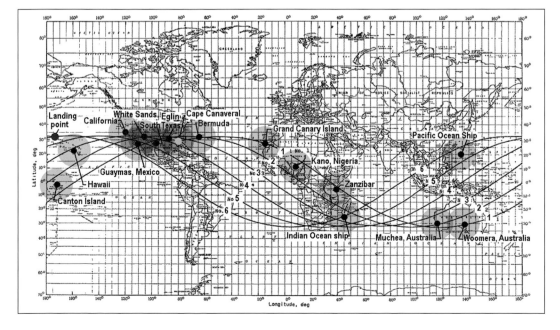

Having managed NASA's High Speed Flight Station at Edwards Air Force Base, California, on 15 September 1959 Walter C. Williams became Associate Director for Project Mercury Operations and eventually established the Operations Coordination Group at Cape Canaveral with Christopher C. Kraft as Flight Director, Stanley White as Chief Flight Surgeon, Merritt Preston as Launch Operations Manager and Scott Simpkinson as Capsule Operations Manager. This group assumed full responsibility for all Mission Control operations. On 9 March they issued titles for the major positions at the MCC, in the launch blockhouse and at the launch pad. Among those soon working in Mission Control was a young Eugene F. Kranz, together with British-born engineer John D. Hodge from Canada and others whose names would became legendary as the space programme evolved.

Cape Canaveral-MCC would control all Mercury and the first three Gemini flights, of which the third (Gemini III) on 23 March 1965 was the first manned flight of the two-seat spacecraft. All subsequent NASA manned flights would be controlled from the MCC relocated to the Manned Spacecraft Center in Houston, Texas, subsequently renamed the Lyndon B. Johnson Space Center. However, Cape Canaveral served as backup to Gemini IV launched on 3 June 1965, the last flight in which that facility was involved.

The legend of 'Mission Control' was built on the back of what was designed and constructed in that building at Cape Canaveral and for the most part it was the men of the 1950s and 1960s that built the reputation that placed America at the centre of world attention. But the greatly unsung contribution that oiled the wheels and sent the vehicles into space on perfect trajectories and data-crunching missions was made by women, and a significant number of those were black and minorities who provided a foundation upon which the greater glory was seized by the male majority.

The cultural stigmas of the age condemned a generation of brilliant, creative women mathematicians and engineers to the backwaters of history simply because they were not supposed to be there – and the written record ignored them. But thankfully we live in more enlightened times and a great debt is owed to the women who contributed to Project Mercury, among whom were people excluded in some communities, outside government control, from sharing the same seat as a white male.

Many more had aspired to join the ranks of astronauts recruited to the US space programme but they were excluded from applications by the way the requirements were written. Several women went through shadow-tests in an attempt to elevate their credentials by some scientists who thought they should be given a chance to represent their country in space. But it was not to be. Mercury astronauts would testify before Congressional committees that in their view women were ill-suited to the demands of space flight and one went so far as to say that women should remain in the kitchen, wear an apron and cook food for when they returned home.

The contributions of the women who served the Mercury programme are given few words in this book because this is not primarily about people, but there are many other sources of detailed information about the contribution they made and the sacrifices given to fulfil their own aspirations to realise inherent skills and help place American men in space. The reader is encouraged to seek out those stories and begin a reading programme based on their efforts as well as those whose names are already legendary, including: *The Mercury 13* by Martha Ackmann (Random House 2004); *Almost Astronauts* by Tanya Lee Stone (Candlewick 2009); *Right Stuff, Wrong Sex* by Margaret A. Weitkamp (Johns Hopkins University Press 2005); and *Promised the Moon* by Stephanie Nolen (Basic Books 2011).

Appendix 2

The astronauts

The selection of astronauts who would fly the Mercury spacecraft began with a written requirement based on what the physicians and the spacecraft engineers believed to be essential criteria. It was also required to conform to President Eisenhower's stipulation that the astronauts selected should come from the military, which resulted in 508 applications submitted through the armed services. The request for applicants was formally published on 22 December 1958, individuals being required to submit their names by 26 January 1959.

Each candidate had to be a US citizen, male, aged 25–39, less than 5ft 11in (180cm) in height, have at least 1,500 hours' flying time, hold at least a bachelor's degree in science or engineering, have had substantial experience in flying research aircraft, possess a tolerance of rigorous or severe environmental conditions and a proven ability to react adequately under stress or emergencies. They were expected to remain with NASA for three years and would receive an initial starting salary of $8,330–$12,770 ($69,375–$106,353 in 2016 money) depending on experience and qualifications.

A down-selection to 110 candidates was made by Stanley C. White, Robert B. Voas and William S. Augerson of whom 58 were from the Air Force, 47 from the Navy and five from the Marine Corps. None of the Army applicants was found suitable. It had originally been thought that an astronaut corps of 12 men was optimal but it became apparent that the calibre of applicants was sufficiently high to believe that none of them would drop out of the rigorous training programme planned so the number was reduced to six.

Of the 69 men screened in Washington six were found to exceed the height restriction; 56 took a range of written tests, were screened in interviews and experienced medical tests and medical history reviews. By March 1958 the group had been reduced to 36 for a much higher level of psychophysiological examination at the Lovelace Clinic in Albuquerque, New Mexico. Four dropped out voluntarily and only one was eliminated from the screening. After a further and still more rigorous series of tests and evaluations, by March just 18 remained, and these were recommended unequivocally. Scrutineers found it impossible to get this total below seven and these were accepted.

The successful candidates were announced to the world on 9 April in a press conference delivered in Washington DC. Together with the

BELOW The Mercury Seven, America's first astronauts. Front row from left: Walter M. Schirra; Donald K. Slayton; John H. Glenn; Malcolm Scott Carpenter. Back row from left: Alan B. Shepard; Virgil I. Grissom; L. Gordon Cooper. *(NASA)*

ABOVE The Mercury Seven's training was the most strenuous of any group selected by NASA, involving fiendish devices to push the human envelope and expand the physiological and psychological profile and capabilities of each man. *(NASA)*

missions they flew or, in the case of Grissom that they were assigned to, these men were:

■ Alan B. Shepard (1923–98), Navy: MR-3; Apollo 14.
■ Virgil I. Grissom (1926–67), Air Force: MR-4; Apollo 1.
■ John H. Glenn (1921–2016), Marines: MA-6; STS-95.
■ Malcolm Scott Carpenter (1925–2013), Navy: MA-7.
■ Walter M. Schirra (1923–2007), Navy: MA-8; Gemini VIA; Apollo 7.
■ Donald K. Slayton (1924–93), Air Force: ASTP.
■ Leroy Gordon Cooper (1927–2004), Air Force: MA-9; Gemini V.

The seven men reported for duty on 27 April 1959 and their training began the following day with a general briefing on the programme, followed next day by briefings on the spacecraft configuration and escape options, support and restraint briefings on 30

ABOVE Jerrie Cobb poses by a Mercury spacecraft mock-up. She was one of the 'Mercury 13', women selected for a private, non-government, physiological screening in parallel with the seven NASA astronauts. This programme was at the instigation of the Lovelace Clinic, which had played a major part in formulating the physical tests for the NASA astronaut candidates. The other members of the Mercury 13 were: Myrtle Cagle, Janet Dietrich, Marion Dietrich, Wally Funk, Sarah Gorelick, Jane Hart, Jean Hixson, Rhea Hurrle, Gene Nora Stumbough, Irene Leverton, Jerri Sloan and Bernice Steadman. *(NASA)*

April and operational concepts and procedures briefings on 1 May. Each astronaut had specific responsibilities for gaining deep insight to selected systems and each would report to the others on what he had learned.

While each man was used to accepting aircraft from manufacturers who had little or no liaison with the pilots who would fly their products, the X-series experimental research aircraft programme had begun to change all that. The X-15 rocket-powered research aircraft benefited greatly from the input on cockpit layout and general ergonomics from North American Aviation test pilot Scott Crossfield and that fed directly to the Mercury programme – although McDonnell design engineers had not been used to having pilots exert quite such a level of authority in such a decisive way! Their contribution to the success of Project Mercury equals that of the engineering design teams.

RIGHT Set up in 1964, the Mercury memorial at the Kennedy Space Center, with Launch Complex 14 in the background. *(NASA)*

Abbreviations and acronyms

A – Amperes (amps).
ABMA – Army Ballistic Missile Agency.
AC – Alternating current.
AFB – Air Force Base.
Ah – Amp hours (battery capacity).
arc-sec – A second of arc.
ARDC – Air Research & Development Command
ARPA – Advanced Research Projects Agency.
ASCS – Automatic Stabilization Control System.
ASIS – Abort Sensing and Implementation System.
BMD – Ballistic Missile Division.
Btu – British thermal units.
Btu/ft^2 – British thermal units per square foot.
Btu/h – British thermal unit hours.
cc/min – Cubic centimetres per minute.
CIA – Central Intelligence Agency.
cm – Centimetres.
cm^3/min – Cubic centimetres per minute.
CO_2 – Carbon dioxide.
CQIS – Coolant quantity indicating system.
db – Decibels.
DC – Direct current.
°/hr – Degrees per hour.
°/sec – Degrees per second.
ECG – Electrocardiogram.
ECS – Environmental Control System.
FBW – Fly-by-wire.
fps – Frames per second.
ft – Feet.
ft/sec – Feet per second.
ft^3 – Cubic feet.
ft^3/min – Cubic feet per minute.
g – Measure of gravitational force.
gm – Grams.
HF – High frequency.
hr – Hours.
H_2O_2 – Hydrogen peroxide.
Hz – Hertz.
ICBM – Intercontinental ballistic missile.

IIP – Instantaneous launch-vehicle impact point.
in^3/min – Cubic inches per minute.
JPL – Jet Propulsion Laboratory.
kC – Kilocoulomb.
kg – Kilograms.
kg/day – Kilograms per day.
kg/hr – Kilograms per hour.
kg/min – Kilograms per minute.
kg/sec – Kilograms per second.
kJ/hr – Kilojoules per hour.
kN – Kilonewtons force.
kN/sec – Kilonewtons per second.
kPa – Kilopascals.
kPag – Kilopascals of gauge pressure.
kph – Kilometres per hour.
kt – Knots.
lb/day – Pounds per day.
lb/hr – Pounds per hour.
lb/in^2 – Pounds per square inch.
lb/in^2g – Pounds per square inch of gauge pressure.
lb/min – Pounds per minute.
lb/sec – Pounds per second.
LES – Launch escape system.
LET – Launch escape tower.
LiOH – Lithium hydroxide.
LJ – Little Joe rocket.
m – Metres.
MA – Mercury Atlas rocket.
MAC – McDonnell Aircraft Company.
MASTIF – Multiple Axis Space Test Inertia Facility.
max-q – Maximum dynamic pressure.
MCC – Mercury Control Center.
M^5 – Man, machine, media, management and mission.
MHz – Megahertz (millions of cycles per second).
min – Minutes.
MISS – Man-In-Space-Soonest.
mm – Millimetres.
mmHg – Millimetres of mercury.
MODM – Manned-One-Day-Mission.
mph – Miles per hour.
MR – Mercury Redstone rocket.

ms – Milliseconds.
MSC – Manned Spacecraft Center.
m/sec – Metres per second.
m^3 – Cubic metres.
m^3/min – Cubic metres per minute.
mV – Millivolts.
N – Newtons force.
NACA – National Advisory Committee for Aeronautics.
NADC – Naval Air Development Center.
NASA – National Aeronautics & Space Administration.
NRO – National Reconnaissance Office.
OAO – Orbiting Astronomical Observatory.
oz – Ounces.
oz/yd^2 – Ounces per square yard.
PARD – Piloted Aircraft Research Division.
RCS – Reaction control system.
RF – Radio frequency.
rpm – Revolutions per minute.
RSCS – Rate Stabilization Control System.
SCORE – Signal Communications by Orbiting Relay Equipment.
sec – Seconds.
SOFAR – Sound fixing and ranging.
SSB – Source Selection Board.
STG – Space Task Group.
UHF – Ultra high frequency.
USAF – United States Air Force.
TAV – Trans-atmospheric vehicle.
V – Volts.
VA – Volt amperes.
VCO – Voltage-controlled oscillator.
VOX – Voice-operated.
W – Watts of electrical energy.
W/C_DA – Weight divided by drag coefficient times the cross-sectional frontal area.
W/cm^2 – Watts per square centimetre.
Wh – Watt hours.
W/in^2 – Watts per square inch.
W/m^2 – Watts per square metre.
WWTN – Worldwide Tracking Network.

Index